Trends in Biomaterials

Trends in Biomaterials

edited by

G. P. Kothiyal
A. Srinivasan

PAN STANFORD PUBLISHING

Published by

Pan Stanford Publishing Pte. Ltd.
Penthouse Level, Suntec Tower 3
8 Temasek Boulevard
Singapore 038988

Email: editorial@panstanford.com
Web: www.panstanford.com

British Library Cataloguing-in-Publication Data
A catalogue record for this book is available from the British Library.

Trends in Biomaterials
Copyright © 2016 by Pan Stanford Publishing Pte. Ltd.
All rights reserved. This book, or parts thereof, may not be reproduced in any form or by any means, electronic or mechanical, including photocopying, recording or any information storage and retrieval system now known or to be invented, without written permission from the publisher.

For photocopying of material in this volume, please pay a copying fee through the Copyright Clearance Center, Inc., 222 Rosewood Drive, Danvers, MA 01923, USA. In this case permission to photocopy is not required from the publisher.

ISBN 978-981-4613-98-9 (Hardcover)
ISBN 978-981-4613-99-6 (eBook)

Printed in the USA

Contents

Preface ix

1. **Bioactive Glass and Glass-Ceramics Containing Iron Oxide: Preparation and Properties** 1
   Nisha Shankhwar, K. Sharma, G. P. Kothiyal, A. Srinivasan
   1.1 Introduction 2
   1.2 Preparation and Processing Techniques 5
       1.2.1 Melt-Quenching Method 5
       1.2.2 Sol–Gel Method 7
       1.2.3 Micro-Emulsion Method 9
       1.2.4 Laser Spinning 10
       1.2.5 Solvent Casting 10
       1.2.6 Gas Phase (Flame Spray) Synthesis 10
       1.2.7 Hydrothermal Synthesis 10
   1.3 Characterization Techniques 11
       1.3.1 Structural Probes 11
       1.3.2 Thermal Analysis 13
       1.3.3 Magnetic Characterization 15
       1.3.4 Surface Analysis 18
             1.3.4.1 Electron spectroscopy for chemical analysis 18
             1.3.4.2 Secondary ion mass spectrometry analysis 18
       1.3.5 Bioactivity Test 19
             1.3.5.1 *In vivo* test 19
             1.3.5.2 *In vitro* test 19
   1.4 Review of Progress in Magnetic Bioglass and Glass-Ceramics 24
       1.4.1 Bioglass Containing Iron Oxide 26
       1.4.2 Magnetic Bioglass-Ceramics 30

2. **Self-Assembly Approach for Biomaterials Development** 49
   Gunjan Verma and P. A. Hassan
   2.1 Phenomena: Principles of Self-Assembly 50
   2.2 Historical Perspective 52

2.3	Design of Self-Assembled Structures		53
2.4	Characterization of Self-Assembled Aggregates		55
	2.4.1	Light Scattering	56
	2.4.2	Neutron Scattering	58
	2.4.3	Small-Angle X-Ray Scattering	59
2.5	Structural Studies on Surfactant Assemblies		60
2.6	Applications		65
	2.6.1	Self-Assemblies for Drug Delivery	65
	2.6.2	Self-Assembly in Biomineralization	71
2.7	Future Directions		76

3. New Trends in Bioactive Glasses: The Importance of Mesostructure — 85

Sevi Murugavel and Chitra Vaid

3.1	Historical Background		86
	3.1.1	Scaffolds for Bone Tissue Engineering	89
	3.1.2	Bioactive Glass Scaffolds	89
	3.1.3	Mesoporous Bioactive Glass Scaffolds	90
3.2	Fabrication of Bioactive Glass and Glass-Ceramic		93
	3.2.1	Melt-Derived Bioactive Glasses	93
	3.2.2	Sol–Gel-Derived Bioactive Glasses	94
		3.2.2.1 Supramolecular chemistry and sol–gel process	94
	3.2.3	Flame Spray Synthesis	95
	3.2.4	Freeze-Drying Technique	96
3.3	Silica-Based Sol–Gel Bioactive Glass Composites		96
3.4	Hybrid Sol–Gel Bioactive Glasses		97
3.5	*In Vitro* Response of Bioactive Mesoporous Glasses		98
3.6	*In Vivo* Response of Bioactive Mesoporous Glasses		101
3.7	Local Structure of MBGs and Glass-Ceramic and its Relation with Dissolution Kinetics		101

4. Biomaterials Based on Natural and Synthetic Polymer Fibers — 121

C. S. Krishna Murthy and Biman B. Mandal

4.1	Introduction	122
4.2	Polymer Fiber Materials	125

	4.2.1	Natural Polymer Fiber Materials	125
		4.2.1.1 Silk	126
		4.2.1.2 Chitin and chitosan	127
		4.2.1.3 Collagen	129
		4.2.1.4 Hyaluronic acid	130
		4.2.1.5 Gelatin	131
		4.2.1.6 Fibrinogen	131
	4.2.2	Synthetic Biodegradable Polymers	132
		4.2.2.1 Absorbable/resorbable polymer fiber materials	132
		4.2.2.2 Polyglycolide	133
		4.2.2.3 Polylactide	134
		4.2.2.4 Poly(ε-caprolactone)	135
		4.2.2.5 Polydioxanone	135
		4.2.2.6 Poly(lactide-co-glycolide)	136
4.3	Fabrication and Processing of Polymer Fibers		137
	4.3.1	Electrospinning	138
		4.3.1.1 Electrospinning apparatus setup	139
		4.3.1.2 Electrospinning processing parameters	140
	4.3.2	Melt Spinning	141
	4.3.3	Wet Spinning	142
	4.3.4	Filament Winding	143
4.4	Characterization, Testing, and Evaluation of Fibers		143
	4.4.1	Characterization of Polymer Fibers	143
	4.4.2	Mechanical Testing and Evaluation	145
		4.4.2.1 Uniaxial tension and burst tests	146
		4.4.2.2 Biaxial tests	149
		4.4.2.3 Identifying fiber orientation	150
4.5	Applications of Polymer Fiber Material		152
	4.5.1	Polymeric Fibers in Biomedical Applications	153
	4.5.2	Extra Cellular Matrix Replacement	153
	4.5.3	Drug Delivery	154
	4.5.4	Wound Dressings	155
	4.5.5	Tissue Engineering Applications	155

		4.5.5.1	Scaffolds for tissue engineering	155
		4.5.5.2	Scaffolds for vascular graft engineering	156
4.6	Conclusion			157

5. Biomaterials in Total Hip Joint Replacements: The Evolution of Basic Concepts, Trends, and Current Limitations—A Review   175

B. Bhaskar, S. Arun, P. S. Rama Sreekanth, and S. Kanagaraj

5.1	Introduction	176
5.2	Components of THR	177
5.3	Evolution of Hip Arthroplasty Technology	179
5.4	Currently Used Biomaterials in Bearing Surfaces	180
	5.4.1 Hard-on-Soft Combinations	180
	5.4.2 Hard-on-Hard Combinations	182
5.5	Biomechanics and Implant Stability	186
5.6	Fixation Techniques	188
	5.6.1 Cemented Fixation	188
	5.6.2 Cementless Fixation	190
	5.6.3 Hybrid Fixation	191
5.7	Current Limitations in THR Surgery	193
	5.7.1 Failure Scenario	193
	5.7.1.1 Accumulated-damage scenario	193
	5.7.1.2 Destructive-wear scenario	193
	5.7.1.3 Particulate-reaction scenario	194
	5.7.1.4 Failed-bonding scenario	194
	5.7.1.5 Stress-shielding scenario	194
	5.7.1.6 Stress-bypass scenario	194
5.8	Biocompatibility	195
5.9	Recent Research and Developments	196
	5.9.1 Reinforcement Technique	196
	5.9.2 Highly Crosslinked Polyethylene Acetabular Cup	196
	5.9.3 Hip Resurfacing Arthroplasty	199
5.10	Conclusion	202

Index   211

Preface

Biomaterials research requires the union of materials scientists, engineers, biologists, biomedical doctors, and surgeons. Societal implications have invoked tremendous interest in this area of research in recent years. It is now possible to tailor make mechanically strong biomaterials, which mimic human bone by controlled crystallization of silico-calcium phosphate glasses with selected additives. Such a bioactive glass-ceramic forms a biologically active hydroxyapatite layer on the surface that permits bonding with bone and soft tissues. Smart metallic alloys have been developed to replace metals as orthopedic implants. What started as a search for strong and durable implant materials has now led to path breaking developments in tissue engineering, targeted drug delivery, and tissue scaffolds. Viable applications of mesoporous structures, polymer biocomposites and fibers (synthetic and natural) in the areas of clinical orthopedics, controlled drug delivery, tissue engineering, orthodontics, etc., have emerged as relatively recent concepts. In the world of bioengineering tissues, the uses of fibers have come a long way as biomaterials due to their various advantages including their robustness, degradation, and their inherent ability to mimic extracellular matrix architecture in three dimensions. This book has attempted to present some of the recent developments in the exciting areas of biomaterials.

The book has five chapters and each chapter has been designed on the lines to give phenomena (physical basis), some historical perspective, materials synthesis and processing, characterization and properties, applications, and future directions to the extent possible. While keeping in mind the above points, a vivid account of the techniques used in the preparation and characterization of these new biomaterials has been brought out. Discussion of the results obtained on the structural, magnetic, antibacterial, bioactivity/compatibility, mechanical, and other related properties and the implication of these results on biomedical applications are given. Each chapter has an exhaustive bibliography for readers to look into more details. The first chapter reviews the progress made in magnetic bioglass and glass-ceramics and their biomedical applications. The second chapter deals with self-assembly approach for biomaterials development. Mesoporous structure and its implications have been brought out in the third chapter. Techniques developed for characterizing the hierarchical structure of sol–gel glasses and hybrids, to quantify the open macroporous networks of scaffolds, have been reviewed. The fourth chapter looks into different categories of natural and synthetic fibers for widespread application as biomaterials. Their specific processing methods and testing and evaluation schemes are thoroughly described along with their advantages and disadvantages. The fifth chapter describes not only an accelerated development and significant advances in biomaterials used in total hip replacement (THR) but also several therapeutic risks that should not be ignored. Technical problems faced by the surgeon during implant fixation in THR and the selection of materials based on the patient's requirement are also discussed. Recent advances in implant materials including the reinforcement and biocompatibility issues have also been addressed.

We thank all the contributors for providing the manuscripts within the stipulated time. We believe that this book will give the readers a fair insight to the recent trends in the area of biomaterials.

G. P. Kothiyal
A. Srinivasan
Winter 2015

Chapter 1

Bioactive Glass and Glass-Ceramics Containing Iron Oxide: Preparation and Properties

Nisha Shankhwar,[a] K. Sharma,[b] G. P. Kothiyal,[c] A. Srinivasan[a]
[a]*Department of Physics, Indian Institute of Technology Guwahati, Guwahati-781039, India*
[b]*Glass and Advanced Materials Division, Bhabha Atomic Research Centre, Mumbai-400085, India*
[c]*Superannuated from Glass and Advanced Ceramics Division, Bhabha Atomic Research Centre, Mumbai-400085, India*
asrini@iitg.ernet.in

The advent of bioactive glass (or simply bioglass) was a major advancement in biomaterials because it mimics human bone in its interaction with human body fluid. The brittle nature of bioglass has led to the development of bioglass-ceramics, which exhibit better mechanical strength. Magnetic bioglasses and bioglass-ceramics have earned a special place in the area of health care, especially in orthopedic applications, cancer therapy, and targeted drug delivery. This chapter reviews the progress made in these important biomaterials, including iron-containing biomaterials and their role in biomedical applications.

1.1 Introduction

Archaeological excavations in Egypt and Eastern Mesopotamia revealed that people made non-transparent glass beads as early as 3500 BC. Before beginning the discussion on bioactive glass and bioactive glass-ceramics (BGCs), it is important to first understand what glass is. Glass is an amorphous (or vitreous) material, which is commonly referred to as a super-cooled liquid. In 1945, the American Society for Testing and Materials defined glass as "an inorganic product of fusion, which has cooled to a rigid condition without crystallizing." If one takes into account that glasses are not restricted to inorganic materials, a more general definition would be "an amorphous solid completely lacking in long range, periodic atomic structure and exhibiting a region of glass transformation behaviour" [1, 2]. Inorganic, organic, or metallic glasses are traditionally formed by rapidly cooling (or quenching) the molten liquid through its melting point or liquidus temperature. Generally, when a molten substance is slowly cooled below its melting point, it undergoes a discontinuous volume change and solidifies into its crystalline (solid) state by liberating heat. However, when a glass-forming melt is cooled below its melting point, no discontinuous volume change is observed and there is no exothermic effect related to the change of the liquid to the solid state. Instead, the viscosity of the melt increases gradually as the temperature is decreased until it attains a value ($\sim 10^{13}$ Poise) typical of a solid.

To understand the formation of the glassy network structure, consider certain inorganic oxides that can form strong ionic/covalent bonds in three-dimensional (3-D) space. This strong bonding presents a kinetic barrier, which prevents bond rearrangement and subsequent crystallization. Silicon dioxide can be regarded as the prototype silicate glass. In pure silica, four strong Si-O-Si links (bridges) exist between SiO_4 tetrahedral units, which result in the formation of a rigid 3-D network. Such network-forming oxides are known as glass formers. Oxides such as Na_2O (referred to as modifier oxides) can react with the network to break these bonds and produce mobile ions, thereby increasing the degree of freedom of the network such that rearrangement becomes easier and making the glass less kinetically stable. Other atoms (referred to as intermediates and modifiers) are linked together with much weaker bonds. Hence, the properties of the glass depend on the exact way in which different types of bonds are distributed in the network. Inorganic compounds, such as oxides, sulfides, and chlorides, consist of positively and negatively charged ions packed together. In melts containing such oxides, the strength of bonds between these oppositely charged ions and their coordination numbers determine the glass-forming ability. Based on the function of the components in glass structures, glass-making oxide components can be divided into three groups:

(1) Glass former (SiO_2, P_2O_5, B_2O_3, GeO_2, As_2O_5, V_2O_5, ZrO_2, Sb_2O_3, etc.), which serves as the backbone of a glass. These oxides can form glass by well-established laboratory techniques.
(2) Modifier (Na_2O, K_2O, CaO, SrO, BaO, etc.), which does not form glass under ordinary conditions but can modify the properties of the glass when used along with a glass former.
(3) Intermediate (Al_2O_3, BeO, ZnO, CdO, PbO, TiO_2, etc.), which occupies a position in between the glass former and modifier.

Silicon dioxide, alkali silicates (Li_2O-SiO_2, Na_2O-SiO_2, K_2O-SiO_2, PbO-SiO_2, SiO_2-Al_2O_3), aluminosilicates, sodium aluminosilicates, calcium aluminosilicates, alkali phosphosilicates, and mineral glasses are some well-known examples of glasses.

Bioactive glass, or simply bioglass, first synthesized in 1969, provided an alternative second-generation implant material for interfacial bonding with host tissue [3–6]. Among various variants of bioglass, the 45S5 (Class A) Bioglass®, originally invented by Hench, is considered to be the benchmark bioglass having a typical composition of $46.1SiO_2 2.6P_2O_5 26.9CaO 24.4Na_2O$ [4–6]. The mechanism of interaction between the bioglass surface and the body fluid, leading to the exchange of ions and the formation of a calcium phosphate surface layer (mimicking the reaction of human bone with body fluid), has been well documented [5–6]. Besides the 45S5 glass, a wide range of bioglass have been developed and are being used in clinical applications. Class A bioactive materials are osteogenetic and osteoconductive, while Class B bioactive materials (e.g., hydroxyapatite) exhibit only osteoconductivity. It has been found that reactions on bioactive glass surface leading to the release of critical concentrations of soluble Si, Ca, P, and Na ions induce intracellular and extracellular responses [5]. For example, a synchronized sequence of genes is activated in osteoblasts that undergo cell division and result in the synthesis of an extracellular matrix, which subsequently mineralizes to become bone. In addition, 45S5 Bioglass has been shown to increase the secretion of vascular endothelial growth factor *in vitro* and enhance vascularization *in vivo*. These characteristic features suggest that scaffolds containing controlled concentrations of bioglass might stimulate neo-vascularization, which is beneficial for building large tissue engineered constructs [7].

The brittle nature of bioglass has posed serious limitations on their use in load-bearing applications. To overcome this hurdle, BGCs have been derived from appropriate bioglass compositions by controlled heat treatment. Hence, BGCs are multi-phase materials that exhibit better mechanical and thermal properties compared to their parent bioglass. Nucleation and growth of bone mineral phases in bioglass are required to obtain BGCs with good bioactivity. Apatite and wollastonite (A-W) BGCs are widely used as implant materials in bone application because of their good bonding ability with bone and ingrowth behavior in bone regeneration.

An impressive progress has been made in developing new materials and refining existing material compositions/microstructure for obtaining better performance in biomedical applications. Biocompatible magnetic materials find a variety of applications such as targeted drug delivery and hyperthermia treatment of cancer. Among the different research activities on bioactive materials related to health care, the sub-areas of developing, processing, and improving bioglass BGCs, and their composites containing magnetic phase(s) have a well-established place. Silica-based nonmagnetic glasses and glass-ceramics with a bioactive glass matrix containing sufficient amount of magnetic phases are potentially useful in hyperthermia treatment of cancer, magnetic-particle-based targeted drug-delivery systems, and magnetic resonance imaging (MRI) [8, 9]. Magnetic hyperthermia treatment uses the heat generated by hysteresis loss in an implanted magnetic material to selectively kill cancerous cells [9–16]. This chapter reviews the advances made in the development of magnetic bioactive glass-ceramics (MgBGCs) mainly for magnetic hyperthermia application. Bioglass and MgBGCs developed by the authors' research groups will be used as examples with special focus on the following systems:

1. $41CaO\text{-}(52\text{-}x)SiO_2\text{-}4P_2O_5\text{-}xFe_2O_3\text{-}3Na_2O$, ($x$ = 0 to 10 mole%), or CSPFN.
2. $4.5MgO\text{-}(45\text{-}x)CaO\text{-}34SiO_2\text{-}16P_2O_5\text{-}0.5CaF2\text{-}xFe_2O_3$ (x = 5 to 20 wt.%) or MCSPC'Fx.
3. $x(ZnO, Fe_2O_3)(65\text{-}x)$ SiO_2 $20(CaO, P_2O_5)$ 15 Na_2O (x = 6 to 21 mole%) with Ca/P = 1.67 and Fe/Zn = 6.5 or ZFSCPN.
4. $2ZnO\text{-}8Fe_2O_3\text{-}25SiO_2\text{-}(50\text{-}x)CaO\text{-}15P_2O_5\text{-}xAg$ (x = 0 to 4 mole%) or ZFSCPAg.
5. $2.07ZnO16.22Fe_2O_319.1SiO2(34.6\text{-}x)CaO27.03P_2O_5\text{-}xHAuCl_4$ (x = 1 to 4 wt.%) or ZFSCPAu.

Traditionally synthesized bulk melt-quenched bioglass and MgBGCs are the major implant materials in use today. However, research and development activities of the last decade show that bulk bioglass and BGCs have several disadvantages. Novel bioceramic materials such as nanocrystalline bioglass and BGCs have much better bioactivity and tissue engineering ability. To familiarize the reader with these recent developments, several popular processing techniques other than the melt-quenching method and some of the salient properties of these novel bioceramics are also discussed in the following sections.

1.2 Preparation and Processing Techniques

1.2.1 Melt-Quenching Method

Glass is prepared by rapidly cooling the molten liquid form of the compound. The cooling rate must be sufficiently fast to preclude crystal nucleation and growth of crystallites. The crystallization rate of an under-cooled liquid depends on the rate of crystal nucleation and on the speed with which the crystal–liquid interface moves. In the melt-quenching method, weighed quantities of the constituent compounds are thoroughly mixed in an agate pestle and mortar or in a ball mill. The well-mixed charge is taken in platinum or alumina crucible and calcined between 600°C to 800°C in an electric furnace to remove extra-gaseous content. In some cases, the calcination process is repeated after regrinding and mixing. After allowing for complete calcination, the temperature of the charge is increased to melt it completely. The molten liquid is held at this temperature for 1–2 h for homogenization. The melt is then poured on a copper plate at room temperature and pressed with another copper plate to quench the molten liquid and form glass samples with thickness of a few millimeters. The as-quenched glass is annealed just below their glass transition temperature (T_g) for a few hours to remove residual stresses accumulated during the quenching process. Glass-ceramics containing bone mineral phases show better biocompatibility and mechanical strength than their parent glass. Controlled heat treatment of as-prepared glass based on the knowledge of the crystallization temperatures of the mineral phases (normally obtained from thermal measurement of glass sample) can yield bioglass-ceramics with desired crystalline phases.

In order to determine the correct mix of raw materials for preparing a glass with a specific composition, a batch calculation is needed. Batch calculation for melt-derived glass with a composition of $44SiO_2$-$41CaO$-$4P_2O_5$-$8Fe_2O_3$-$3Na_2O$ is illustrated as follows (also see Table 1.1). Source compounds taken for CaO, P_2O_5, and Na_2O are $CaCO_3$, $(NH_4)_2HPO_4$, and $Na_2CO_3 \cdot H_2O$, respectively.

Table 1.1 Batch calculation for preparing glass by the melt-quenching route

Source compound (MW g/mol)	Batch compound (MW g/mol)	Wt. fraction (B)	Gravimetric factor (C)	B × C (g)
SiO_2 (60.08430)	SiO_2 (60.08430)	0.379	1	0.379
CaO (56.07740)	$CaCO_3$ (100.0869)	0.330	1.79	0.591
P_2O_5 (141.9445)	$(NH_4)_2HPO_4$ (132.06)	0.081	0.93	0.075
Fe_2O_3 (159.6882)	Fe_2O_3 (159.6882)	0.183	1	0.183
Na_2O (61.97894)	$Na_2CO_3 \cdot H_2O$ (124.0037)	0.027	2	0.054

- **Mol. wt. (MW) of glass (A)** = [(MW of SiO$_2$ × 44) + (MW of CaO × 41) + (MW of P$_2$O$_5$ × 4) + [(MW of Fe$_2$O$_3$ × 8) + (MW of Na$_2$O × 3)]/100 = 69.741 g/mol.
- **Wt. fraction of source compound (B)** = (wt. % of source compound × MW of source compound/MW of glass (A) × 100.
- **Gravimetric factor (C)** = MW of each batch compound/MW of corresponding source component.
- Batch compounds for 1 g of 44SiO$_2$41CaO4P$_2$O$_5$8Fe$_2$O$_3$3Na$_2$O glass = (B) × (C)

1.2.2 Sol–Gel Method

The synthesis of silica-based bioactive glass by the sol–gel technique at low temperatures using metal alkoxides, e.g., tetra ethyl ortho silicate (TEOS), as precursors was first demonstrated in 1991 [17]. The sol–gel process involves the following steps:

(1) Hydrolysis of an appropriate metal alkoxide under acidic or basic environment.
(2) Addition of network modifiers and other additives to the solution to obtain the desired product composition.
(3) Gelation of the solution. Ageing is required to ensure complete gelation.
(4) Freeze drying of the gel to obtain a dry powder.
(5) Calcination of the powder to remove extra-gaseous substances.

Depending on the procedure adopted, sol–gel-derived products may be 2-D (thin film) or 0-D (nanoparticle) nanomaterial with a large specific surface area and a high degree of porosity. Apart from the specific surface area and the pore size, surface morphology plays an important role in influencing the *in vitro* bioactivity of nanobioglass particles. A variety of metal ions such as zinc, magnesium, zirconium, titanium, boron, and silver have been added to bioactive glass for enhancing the glass functionality and bioactivity [18–23]. However, literature shows that fully amorphous bioactive glass is extremely difficult to synthesize by the sol–gel route because of the spontaneous precipitation of various precursors during polymerization and gelation steps, which is very difficult to control. However, it has been demonstrated that it is possible to obtain fully amorphous 45S5 glass composition by the sol–gel route [24]. A flowchart explaining the steps involved in the preparation of 45S5 Bioglass by the sol–gel route is given in Fig. 1.1. Batch calculation for sol–gel-derived 45S5 Bioglass with a composition of 45S5 45SiO$_2$ 24.5CaO 24.5Na$_2$O 6P$_2$O$_5$ is given in Table 1.2 to demonstrate the procedure for another system with different source and batch components.

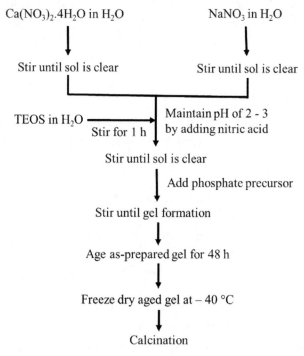

Figure 1.1 Sol–gel procedure for preparing 45S5 glass.

Table 1.2 Batch calculation for preparing 45S5 Bioglass by the sol–gel route

Source component (MW g/mol)	Batch compound (MW g/mol)	Wt. fraction (B)	Gr. factor (C)	B × C (g)
SiO$_2$ (60.08430)	TEOS (208.33)	0.419	3.47	1.454
Na$_2$O (61.97894)	NaNO$_3$ (84.99)	0.234	1.37	0.320
CaO (56.07740)	CaNO$_3$·4H$_2$O (236.15)	0.213	4.21	0.897
P$_2$O$_5$ (141.9445)	(NH$_4$)$_2$HPO$_4$ (132.06)	0.130	0.93	0.123

Molecular weight of glass = [(MW of SiO$_2$ × 45) + (MW of Na$_2$O × 24.5) + (MW of CaO × 24.5) + (MW of P$_2$O$_5$ × 6)]/100 = 64.478 g/mol

1.2.3 Micro-Emulsion Method

A micro-emulsion is a thermodynamically stable, transparent, and isotropic dispersion of two immiscible liquids such as water and oil stabilized by surfactant molecules at the water–oil interface. In water-in-oil micro-emulsions, nanosized water droplets are dispersed in the continuous hydrocarbon phase and surrounded by a monolayer of surfactant molecules. The aqueous droplet diameter is usually in the range 5–20 nm. These aqueous droplets act as a nanoreactor inside which reactions can take place when the droplets containing the suitable reactants collide with each other. To achieve this, the precursor particles of hydroxide or oxalate are first formed in a micro-emulsion system. After drying and calcination of the precursor powder at an appropriate temperature, the desired oxide system is obtained. SiO$_2$-CaO-P$_2$O$_5$ bioglass nanoparticles have been synthesized by the micro-emulsion method [25]. The typical recipe for this process starts with the addition of TEOS, which is used as a silica precursor to 25% aqueous NH$_4$OH, drop-wise under constant stirring until TEOS undergoes complete hydrolysis. Then the phosphate precursor (say, triethylphosphate) is added to this solution followed by the CaO precursor [say, aqueous 0.2 M Ca(NO$_3$)$_2$·4H$_2$O]. After ageing at room temperature, the white precipitate obtained is collected by centrifugation, washed with absolute ethanol, and then dried at 50°C in a vacuum oven for a day. The dry powder has to be calcined at 600°C to remove the oil and surfactants and yield the SiO$_2$-CaO-P$_2$O$_5$ glass nanoparticles with size in the range of 25–50 nm. The diameter of the nanoparticles obtained by such a procedure was found to be dependent on the molar ratio of water to surfactant in water–oil emulsions. The size of water droplets was also found to increase with an increase in the molar ratio of water to surfactant [26].

Apart from the above-mentioned popular techniques for preparing bioglass and BGC, the following techniques have also been employed.

1.2.4 Laser Spinning

Large quantities of nanofibers can be produced using this method with specific and controllable chemical compositions without the necessity of any chemical additives or post-preparative heat treatments. The laser spinning technique essentially involves quick heating and melting of a small volume of the precursor material using a high power laser. A supersonic gas jet injected into the melt volume quickly stretches and cools the molten material. Long glass fibers with high length-to-diameter ratios have been produced by the elongation of the viscous molten material. The product is in amorphous form because of the high cooling speed. 45S5 and 52S4.6 Bioglass nanofibers have been prepared by this method [27].

1.2.5 Solvent Casting

Solvent casting of the composite scaffolds involves the dissolution of a polymer in an organic solvent, mixing with bioactive ceramic or glass granules, and casting the solution in a predefined 3-D mold [28]. The solvent is subsequently allowed to evaporate. The main advantage of this technique is the ease of fabrication without the need of any specialized equipment. 45S5 Bioglass–poly(D,L-lactide) composite has been prepared by solvent casting [29].

1.2.6 Gas Phase (Flame Spray) Synthesis

Gas phase synthesis uses metal–organic precursor compounds to produce nanoparticles at temperatures above 1000°C. The basic principle of all gas phase synthesis methods is the formation of molecular nuclei followed by condensation and coalescence of the nuclei, thereby leading to the growth of nanoparticles at high temperatures. Flame spray synthesis, a technique known since 1940 [30], is now used to produce megatons of silica and titania nanoparticles per year.

1.2.7 Hydrothermal Synthesis

Hydrothermal synthesis includes the crystallization of materials from high-temperature aqueous solutions at high pressure. The process utilizes single or heterogeneous phase reactions in aqueous media at elevated temperature ($T > 25°C$) and pressure ($P > 100$ kPa) to crystallize ceramic materials directly from a solution [31]. The synthesis depends on the solubility of materials in hot water at high pressure, and the crystal growth takes place in an autoclave. Hydrothermal synthesis offers many advantages over conventional and non-conventional ceramic synthetic methods as different forms of ceramics can be prepared with hydrothermal synthesis, such as powders, fibers, single crystals, monolithic ceramic bodies, and coatings on metals, polymers, and ceramics. This technique has been employed in the synthesis of nanorods of bioceramics (hydroxyapatite), among others [32].

1.3 Characterization Techniques

The most prominent characterization techniques available for determining the structural, thermal, magnetic, and surface properties of glass and glass-ceramics are discussed in the following subsections.

1.3.1 Structural Probes

X-ray diffraction (XRD) is the primary structural probe used for verifying the amorphous nature of glass samples and to study the phase purity and crystal structure of glass-ceramics. Characteristic X-rays emitted by Cu, Mo, etc. have wavelengths (λ) of the order of interatomic distances in solids. A periodic arrangement of atoms as in crystals can diffract these X-rays whenever the Bragg condition for constructive interference of X-rays reflected from consecutive layers of a plane is satisfied [33].

These reflections and their angular positions serve as fingerprints to identify the crystal structure. In most cases, the XRD pattern obtained for a given crystalline compound is compared with a large database of crystal structures, such as the International Center for Diffraction Data, using search and match software. XRD patterns can be refined by performing least squares fitting to identified crystal structure by the Rietveld method to obtain a plethora of structural parameters such as bond lengths, bond angles, phase purity, etc.

[34]. When the number of layers in a plane is considerably reduced as in nanocrystalline materials, the reflections appear broadened [33, 35]. The average size (d) of the fine crystallites can be estimated from the broadened peak using Scherrer's formula:

$$d = \frac{k\lambda}{\beta \cos\theta} \quad (1.1)$$

where β is the full width at half maximum of the Bragg peak, θ is the diffraction angle at the peak maximum, and k is a constant, which depends on the shape of the crystallite. The Williamson–Hall method [36] attributes peak broadening to both microstrain as well as crystallite size and enables the estimate of both these parameters.

On the other hand, the XRD pattern of amorphous materials such as glass do not exhibit any sharp reflections but characteristic humps as shown in Fig. 1.2a [37]. When glass is heat treated above its glass transition temperature, crystallization is initiated. Under such conditions, bone mineral phases and magnetite phases in nanocrystalline form are obtained as depicted in Fig. 1.2b [38]. Powder neutron diffraction (ND) [39] and electron diffraction [40] are also used for structural characterization of glass and glass-ceramics. Since neutrons carry a magnetic moment, ND can provide information on the magnetic structure of materials apart from enabling isotope substitution studies, which are not possible with the other diffraction techniques. However, the need for a nuclear reactor or a spallation source restricts the accessibility of this technique to most users. Electron microscopes, such as transmission electron microscope (TEM) and scanning electron microscope (SEM), are more easily available. However, the TEM technique requires careful sample preparation to provide thin sections to act as transmission grating for electrons. An SEM is an instrument used to observe the morphology of a sample at higher magnification, resolution, and depth of focus compared to an optical microscope. Herein an accelerated beam of mono-energetic electrons is focused onto the surface of the sample and is scanned over a small area. The back-scattered or secondary electrons emanating from the sample surface/bulk give the signature of microstructure. For analysis, the sample surface should be electrically conducting to prevent charge build-up. This is generally achieved through gold coating where a thin layer of gold is sputtered on the sample surface to render it electrically conducting.

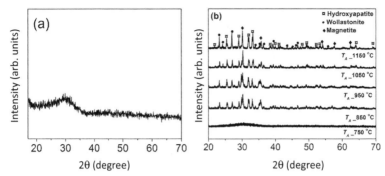

Figure 1.2 XRD pattern of CaO-SiO$_2$-P$_2$O$_5$-Na$_2$O-Fe$_2$O$_3$ (a) glass and (b) glass-ceramics derived after heat treatment at different temperatures T_A. Reprinted with permission from Ref. 38, Copyright 2014, IEEE.

1.3.2 Thermal Analysis

Various techniques are available for studying the thermal behavior of glass and glass-ceramics, such as differential thermal analyzer (DTA), differential scanning calorimeter (DSC), thermal gravimetric analyzer (TGA), and thermo-mechanical analyzer (TMA). With DTA and DSC techniques, the thermal behavior of a sample can be studied over a wide temperature range, under isothermal and non-isothermal conditions. In these techniques, the sample and a reference material are maintained at the same temperature with respect to each other. In a DTA, the excess thermal energy between the sample and the reference is measured in terms of ΔT as a function of temperature or time. In a DSC, the heat flow (dH/dt) to the sample with respect to a reference at the same temperature is recorded as a function of temperature or time. Although DSC is more sensitive than DTA, the higher cost of DSC with operating temperatures above 600°C and the availability of units that can simultaneously measure ΔT as well as weight loss of a sample make DTA a popular instrument. The reference used in a DTA/DSC is an inert material such as alumina or just an empty aluminum sample pan.

In non-isothermal measurements, the temperatures of the sample and the reference are increased at a constant heating rate. Thermal changes in a sample may be exothermic or endothermic. Some of the endothermic transitions are glass transition, reduction, dehydration, and some decomposition reactions. Crystal–crystal structure transformation, glass-crystal transition, oxidation, and some decomposition reactions are the most frequently studied exothermic transitions. As glass is heated at a constant heating rate in a DSC, heat flow exhibits an endothermic baseline shift at the glass transition temperature (T_g), followed by an exothermic crystallization peak at T_c. Further heating melts the sample at T_m, which is an endothermic transformation. Generally, multi-component glasses show one or more exothermic peaks corresponding to the de-vitrification of various crystalline phases (see Fig. 1.3). Such glasses may sometimes show multiple melting endotherms as well. In principle, a TGA is a weighing balance with the sample suspended inside a furnace so that the mass loss in a sample can be recorded as a function of increasing temperature (with constant heating rate), or as a function of time. A TGA is commonly used to determine selected characteristics of materials that exhibit either mass loss or mass gain ($\pm\Delta m$) due to decomposition, oxidation, or loss of volatiles (such as moisture) [41].

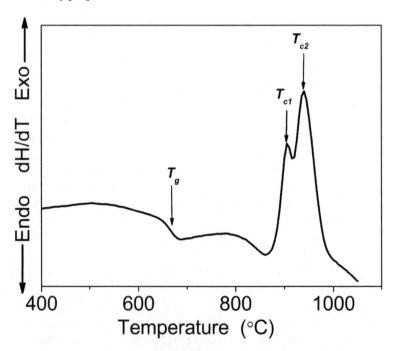

Figure 1.3   DSC curve recorded for a CaO-SiO_2-P_2O_5-Na_2O-Fe_2O_3 glass at a heating rate of 20°C/min.

Commercial units combining both DTA and TGA enable simultaneous recording of ΔT and Δm as a function of temperature or time [42]. On the other hand, a TMA records a change in the dimension (dilation) or a mechanical property of the sample while it is subjected to a temperature variation. The effect of temperature modulation with a set frequency and amplitude on a sample can also be monitored using this instrument [43].

1.3.3 Magnetic Characterization

Various techniques are available for magnetic characterization, such as superconducting quantum interference device (SQUID) based magnetometer, vibrating sample magnetometer (VSM), and electron paramagnetic resonance (EPR) spectrometer. Although the working principles behind a SQUID magnetometer and a VSM are different, they ultimately measure the same magnetic parameters. Sensitivity and applicable magnetic fields are much higher in SQUID magnetometers. However, electromagnet-based VSMs are cost-effective solutions to low-field (0–20 kOe) studies. A VSM can measure the DC magnetic moment as a function of temperature (T), magnetic field (H), angle, and time. So it allows performing magnetic susceptibility (χ), $(M\text{-}H)_T$, and $(M\text{-}T)_H$ studies. The induction heat generated by a thermoseed subjected to an alternating magnetic field of frequency f is a product of f and the magnetic hysteresis loop area. Measurements of magnetic moments as small as 5×10^{-5} emu are possible in magnetic fields from zero to 20 kOe. Figure 1.4 shows the evolution of magnetism in a bioglass composition heat-treated at different temperatures (T_A). It would be informative to compare the M-H data obtained at each T_A with the corresponding XRD data in Fig. 1.2b.

Iron exists in Fe^{2+} or Fe^{3+} ionic states in glass and glass-ceramic samples containing iron oxide, and the interaction between these ions determines the magnetic property of the sample. Super-exchange interaction among the iron ions in oxide glass has been mostly attributed to an antiferromagnetic coupling within the pairs Fe^{3+}-Fe^{2+}, Fe^{3+}-Fe^{3+}, and Fe^{2+}-Fe^{2+} [44–47]. Negative values of a parameter called paramagnetic Curie temperature (θ_p) indicate the existence of super-exchange magnetic interactions in glass between the iron ions, which are mostly antiferromagnetically coupled. θ_p can be estimated from the relation [48]:

$$\chi^{-1} = (T - \theta_p)/C_M \quad (1.2)$$

where χ^{-1} is the inverse magnetic susceptibility and C_M is the Curie constant. In actual practice, $\chi^{-1}(T)$ data are obtained well inside the paramagnetic regime. According to Eq. 1.2, these data should exhibit a linear Curie–Weiss type of behavior. θ_p can be estimated from least squares linear fit to the data. The molar fraction of Fe^{3+} (say, x_1) and Fe^{2+} (say, x_2) ions in the glass can be calculated using the relations:

$$x\mu_{exp}^2 = x_1\mu_{Fe^{3+}}^2 + x_2\mu_{Fe^{2+}}^2 \quad (1.3)$$

$$x = x_1 + x_2 \quad (1.4)$$

where $\mu_{exp} = 2.827(C_M/2x)^{1/2}$, $\mu_{Fe^{2+}} = 4.90\ \mu B$, and $\mu_{Fe^{3+}} = 5.92\ \mu B$.

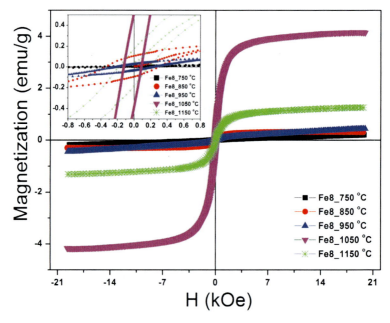

Figure 1.4 *M-H* loops of a CaO-SiO$_2$-P$_2$O$_5$-Na$_2$O-Fe$_2$O$_3$ glass heat treated at different temperatures. Inset gives an enlarged view of data close to the origin. Reprinted with permission from Ref. 38, Copyright 2014, IEEE.

EPR spectroscopy studies of glass/glass-ceramics containing magnetic ions reveal the nature of interactions between the magnetic ions apart from providing information on the valence state of ions present. EPR spectroscopy studies chemical species that have one or more unpaired electrons, such as organic and inorganic free radicals or inorganic complexes containing a transition metal ion. Every electron has a magnetic moment and spin quantum number $s = 1/2$, with magnetic components $m_s = \pm 1/2$. When an external magnetic field of strength B_0 is applied, the electron's magnetic moment aligns itself either parallel ($m_s = -1/2$) or antiparallel ($m_s = +1/2$) to the field.

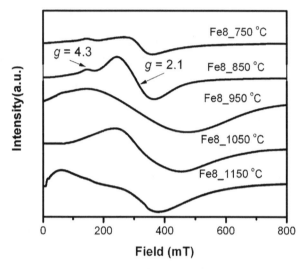

Figure 1.5 EPR spectra of a CaO-SiO$_2$-P$_2$O$_5$-Na$_2$O-Fe$_2$O$_3$ glass heat treated at different T_A. Suppression of $g = 4.3$ resonance for samples annealed at $T_A > 850°C$ shows the formation of Fe clusters at higher T_A. Reprinted with permission from Ref. 38, Copyright 2014, IEEE.

Parallel alignment corresponds to the lower energy state, and separation between it and the upper state is $\Delta E = g_e \mu_B B_0$, where g_e is the electron's so-called g-factor and μ_B is the Bohr magneton. An unpaired electron can move between the two energy levels by either absorbing or emitting electromagnetic radiation of energy $\varepsilon = h\nu$ such that the resonance condition $\varepsilon = \Delta E$ is obeyed. This leads to the fundamental equation of EPR spectroscopy: $h\nu = g_e \mu_B B_0$. Experimentally, this equation permits a large combination of frequency and magnetic field values, but the great majority of EPR measurements are made with microwaves in the 9–10 GHz region, with fields corresponding to about 0.8 T. EPR spectra can be generated by either varying the photon frequency incident on a sample while holding the magnetic field constant, or doing the reverse. In practice, the frequency is usually kept fixed (Fig. 1.5). A collection of paramagnetic centers, such as free radicals, is exposed to microwaves at a fixed frequency. By increasing an external magnetic field, the gap between the $m_s = \pm 1/2$ energy states is widened until it matches the energy of the microwaves. At this point, the unpaired electrons can move between their two spin states. Since there are more electrons in the lower state due to the Maxwell–Boltzmann distribution, there is a net absorption of energy, and this absorption is monitored and converted into a spectrum.

1.3.4 Surface Analysis

1.3.4.1 Electron spectroscopy for chemical analysis

Electron spectroscopy for chemical analysis (ESCA) or X-ray photo electron spectroscopy (XPS) technique provides information about the elemental oxidation states and can be used to investigate the structural changes in glass [49]. It is a surface-sensitive technique that measures the elemental composition and can be used to study the surface modification on implant surfaces when immersed in physiological environment.

1.3.4.2 Secondary ion mass spectrometry analysis

Secondary ion mass spectrometry (SIMS) is another technique used to analyze surface composition. This technique involves bombardment of the specimen with a focused primary ion beam and collection and analysis of the ejected secondary ions [50]. The mass–charge ratios of the secondary ions are measured with a mass spectrometer to determine the elemental composition of the surface up to a depth of 1 to 2 nm. Since SIMS measures the mass of emitted ions, it is useful for detecting low-mass elements like hydrogen or deuterium. In addition, it can provide the depth profiles of biomaterial surfaces kept under physiological environment [51].

1.3.5 Bioactivity Test

From biological point of view, biocompatibility relates to the acceptability of synthetic biomaterials in the living body (especially of mammals/humans). A bioactive implant material is expected to evoke the same response as bone in physiological condition, i.e., the formation of a calcium phosphate or more appropriately a hydroxyapatite (HA)] surface layer. Biocompatibility testing and evaluation involve both *in vitro* and *in vivo* assessments.

1.3.5.1 *In vivo* test

In vivo testing of biomaterials involves small animal models such as mouse, rat, or rabbit. *In vivo* tests provide interactions with extracellular matrix, bone marrow, soft tissue, center of bones where blood cells are made, protein, and molecules. *In vivo* assays use animal models to test the biological response to biomaterials in living animals with active fully integrated systems. This response provides a dual evaluation based on acquiring clinical data during the trial period and histological data from the examination of sections of post-mortem tissue to assess the tissue response to implanted materials.

1.3.5.2 *In vitro* test

To reproduce the formation of HA layer on bioactive materials *in vitro*, Kukubo and his colleagues developed an acellular (modified) simulated body fluid (SBF) that has the same inorganic ion concentrations as human blood plasma (HBP) (see Tables 1.3 and 1.4) [52, 53]. This fluid can be used not only for the evaluation of the bioactivity of artificial materials *in vitro*, but also for coating apatite layer on various materials under biomimetic conditions.

In vitro bioactivity test involves immersion of plate-shaped samples in SBF at 36.5°C and inspecting the sample surface for evidence of formation of HA layer after several days. Surface structural changes on the samples treated in SBF are usually analyzed by grazing incidence X-ray powder diffraction (GI-XRD), Fourier transform infrared (FT-IR) spectroscopy, scanning electron microscopy (SEM), and energy dispersive X-ray spectroscopy (EDS) techniques.

Table 1.3 Ion concentrations in SBF and HBF

Ion (mM)	Na⁺	K⁺	Mg²⁺	Ca²⁺	Cl⁻	HCO₃⁻	HPO₄²⁻	SO₄²⁻
SBF	142	5	1.5	2.5	147.8	4.2	1	0.5
HBP	142	5	1.5	2.5	103	2.7	1	0.5

Table 1.4 Reagents and their order of addition for preparing modified SBF

Order	Reagent	Amount	Formula weight
1	NaCl	8.035 g	58.4430
2	NaHCO₃	0.355 g	84.0068
3	KCl	0.225 g	74.5515
4	K₂HPO₄·3H₂O	0.231 g	228.2220
5	MgCl₂·6H₂O	0.311 g	203.3034
6	1.0M HCl	39 ml	—
7	CaCl₂	0.292 g	110.9848
8	Na₂SO₄	0.072 g	142.0428
9	Tris buffer	6.118 g	121.1356
10	1.0M HCl	0–5 ml	—

Note: Item 10 is added at the end to adjust the pH [53].

In GI-XRD studies, a well-collimated beam of monochromatic X-rays is used to investigate the structure of atomic layers close to the surface by controlling the depth of penetration of the X-rays by maintaining low (grazing) incident angles. This is a commonly used method for structural evaluation of surface layers or thin films. When a bioactive material is immersed in SBF for a day or more, a surface layer forms on its surface. The mechanism of the formation of the HA surface layer of bioglass immersed in SBF is well documented [52]. Figure 1.6 shows the GI-XRD pattern obtained for a CaO-SiO_2-P_2O_5-Na_2O-Fe_2O_3 glass treated in SBF for several days. It can be seen that the XRD pattern of the untreated glass (designated as 0d) shows the amorphous features of the glass. With increase in immersion time, first an amorphous surface layer forms (as seen in the XRD pattern designated as 1d), followed by partial and then fully crystallized HA layer after 30 days of immersion. The broad reflections indicate that the surface layer is made of nanocrystalline HA. GI-XRD data also reveal the preferred orientation of the HA crystallites. FT-IR provides a fingerprint of the functional groups in a material in the form of characteristic infrared bands corresponding to their vibrational modes.

Figure 1.7 illustrates the FT-IR spectra of the same SBF-treated CaO-SiO$_2$-P$_2$O$_5$-Na$_2$O-Fe$_2$O$_3$ glass whose GI-XRD patterns are shown in Fig. 1.6. The evolution of C-O, P-O, and P-O-P absorption bands provides insight into the nature of growth of the HA surface layer with immersion time in SBF. The surface morphology of a bioactive glass treated in SBF can be characterized by SEM. The micrographs provide visual evidence of the formation of a surface layer. In scanning or field emission electron microscope, the most common or standard detection mode is the secondary electron imaging.

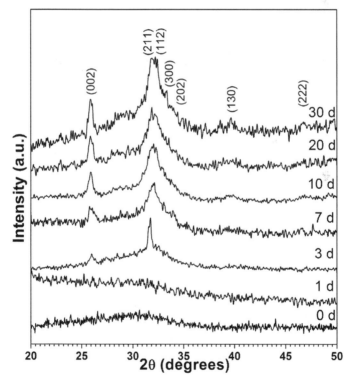

Figure 1.6 GI-XRD patterns of a CaO-SiO$_2$-P$_2$O$_5$-Na$_2$O-Fe$_2$O$_3$ glass sample soaked in SBF for various periods. Reprinted from Ref. 54, Copyright 2009, with permission from Elsevier.

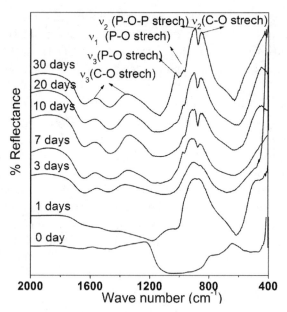

Figure 1.7 FT-IR spectra of a CaO-SiO$_2$-P$_2$O$_5$-Na$_2$O-Fe$_2$O$_3$ glass treated in SBF for various days. Reprinted from Ref. 54, Copyright 2009, with permission from Elsevier.

SEM images of the surfaces of MCSPC'F samples soaked in SBF for different time are shown in Fig. 1.8. The growth of the HA layer with immersion time in SBF can be seen. SEMs equipped with EDS can be used to identify the composition of this surface layer. XPS spectra of MCSPC'F samples before immersion in SBF are shown in Fig. 1.9a. The electronic density of the bonding states of the silicon and oxygen atoms gets modified as Fe$_2$O$_3$ is added to the silica network [55]. This results in the formation of non-bridging oxygen. This shift in core level transitions in an XPS spectrum is sensitive to oxidation states. The difference between bridging and non-bridging oxygen atoms is reflected in the XPS chemical shift of the O1s peak. The XPS technique is used to identify the contributions to the O1s core level spectrum from the bridging and non-bridging oxygen atoms. O1s photoelectron plot for the samples having different amounts of iron oxide is shown in Fig. 1.9b. Along with the curve-fitting procedure, the low binding energy peak at ~530 eV corresponds to the contribution from the non-bridging oxygen atoms, while the peak at ~531 eV is attributed to the bridging oxygen. The areas under the two peaks (non-bridging and bridging) will give the fraction of non-bridging oxygen. XPS scans of MCSPC′F samples after immersion in SBF are shown in Fig. 1.9c. It can be seen that the intensities of peaks

Figure 1.8 SEM images of the surfaces of MCSPC′F samples after (a) 1 week and (b) 4 weeks immersion in SBF. Reprinted from Ref. 56, Copyright 2009, with permission from Elsevier.

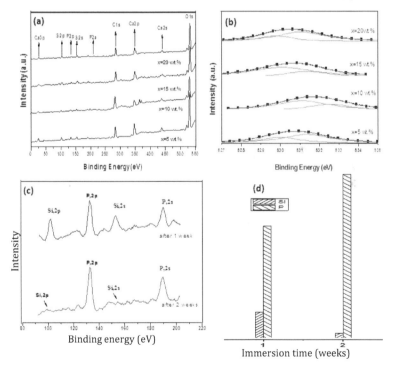

Figure 1.9 (a) XPS of MCSPC′F samples, (b) magnified view of O1s peak with fitting for NBO and BO, (c) magnified view for Si and P peaks, and (d) elemental concentration of the same after one and two weeks of immersion in SBF. Reprinted from Ref. 56, Copyright 2009, with permission from Elsevier.

corresponding to Si and P vary with an increase in sample immersion time. Increase in P content and a net decrease in Si on the sample have been observed (Fig. 1.9d), which are due to the adsorption of phosphate ions from the SBF solution.

The SIMS spectra of MCSPC'F samples after immersion in bovine serum albumin (BSA) are shown in Fig. 1.10. Ca (m/z = 40) and Fe (m/z = 56) are observed as major peaks, while some peaks corresponding to C and Si are also visible. The peaks corresponding to CH_3, CHN, CO, $CONH_2$, $C_2H_5NH_2$, CN_3H_3, C_2H_6NO, $C_2H_2CONH_2$, and $C_3H_4CONH_2$, which are related to the adsorbed BSA, are evident at m/z = 15, 27, 28, 44, 45, 57, 60, 70, and 84, respectively. The spectra of sputtered surfaces show that as sputtering time is increased, the peaks corresponding to polymeric chain are removed, indicating the surface coverage with polymeric layer.

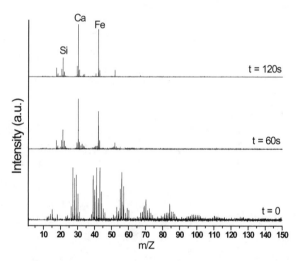

Figure 1.10 SIMS data of MCSPC'F samples obtained for different sputtering times. Reprinted from Ref. 51, Copyright 2012, with permission from Elsevier.

1.4 Review of Progress in Magnetic Bioglass and Glass-Ceramics

Prior to the arrival of bioglass in 1969, all available implant materials such as metals and polymers were designed to be bioinert. However, these materials triggered fibrous encapsulation after implantation, rather than forming a stable interface or bond with tissues. Professor Hench's efforts launched the field of bioactive ceramics, with many new materials and products being formed from variations on bioactive glass [3–5] and ceramics such as synthetic HA and other calcium phosphates [3, 4, 57] that stimulate a beneficial response from the body, particularly bonding to host tissue (usually bone). Applications of bioactive materials in health care such as implants and fillers are well known [57, 58]. On the other hand, applications such as thermoseeds in hyperthermia treatment of cancer and magnetic-particle-based targeted drug-delivery systems require bioactive materials with magnetic content. Hyperthermia treatment of cancer is based on the observation that cancer cells survive less than ordinary tissue cells when heated to 42–45°C. When the part of the human containing the tumors is subjected to local heating to temperatures above 42°C, the tumors are easily heated than the surrounding healthy tissues since blood vessels and nervous systems in the tumors are poorly developed. So cancer cells are easily killed by this heat treatment since oxygen supply via blood vessels is not sufficient in the tumor [8–16, 37]. Various methods such as treatment with hot water, infrared rays, ultrasound and microwaves have been attempted for heating tumors. The concept of utilizing magnetic materials and an alternating magnetic field for this treatment has been proposed by many researchers [9–10]. When magnetic particles implanted around a bone tumor are placed under an alternating magnetic field, they will heat up the locality due to magnetic hysteresis loss. Calibrated application of the alternating magnetic field can effectively and selectively kill the cancerous cells in the locality. Localized hyperthermic treatment for musculoskeletal tumors with ferromagnetic ceramics was first reported by Kokubo *et al.* [59] using an animal model. Their work showed that a bioactive and (ferro) magnetic glass-ceramic obtained by the heat treatment of Fe_2O_3-CaO·SiO_2-B_2O_3-P_2O_5 glass can be used as a thermoseed [60]. Two patents on this topic (U.S. Patent 4323056 entitled "Radio frequency induced hyperthermia for tumour therapy" filed by N.F. Borrelli *et al.* in 1982 and U.S. Patent 6167313 entitled "Targeted hysteresis hyperthermia as a method for treating diseased tissue" filed by B.N. Gray and S.K. Jones in 2000) need special mention here. Before reviewing the developments in MgBGCs, it is informative to look at the progress made in the parent bioactive glass containing iron oxide.

1.4.1 Bioglass Containing Iron Oxide

Silicate glass is the most preferred bioactive glass because glass with silica content higher that 50% exhibits excellent bioactivity [61]. In the attempt to prepare parent bioglass with iron oxide, glass compositions close to the 45S5 Bioglass such as $41CaO(52-x)SiO_2 4P_2O_5 xFe_2O_3 3Na_2O$ ($x = 0$ to 10 mole%) (CSPFN) happen to be a natural choice. Methods to prepare CSPFN glass include melt-quenching [37] and sol–gel [24] techniques. Melt-quenched CSPFN glass has been studied in detail [37]. T_g was found to decrease from 656°C to 605°C when iron oxide concentration was increased from 2 to 10 mole%, which has been attributed to the de-polymerization of the glass skeleton with increase in iron oxide content. DSC studies showed that this glass exhibited two crystallization temperatures (T_c). The first is due to the crystallization of HA and the second one has been attributed to wollastonite. T_c's of both the crystalline phases decrease with increasing iron oxide concentration in glassy matrix from 2–10 mole% [37]. In earlier studies also, a similar trend was observed [58]. DSC studies on A/W glass containing iron oxide, viz., $4.5MgO(45-x)CaO 34SiO_2\ 16P_2O_5 0.5CaF_2 xFe_2O_3$ ($x = 0$ to 20 wt.%) (MCSPC'F) glass [37], showed similar variations in T_g and T_c's. $x(ZnO, Fe_2O_3)(65-x)SiO_2\ 20(CaO, P_2O_5)15\ Na_2O$ ($x = 6$ to 21 mole%) (ZFSCPN) glass with optimized Ca–P (~1.67) and Fe–Zn (~6.5) ratios has also been synthesized by the melt-quenching route [37]. The choice of the ions ratios corresponds to the ratio in HA and $Zn_{0.4}Fe_{2.6}O_4$. DSC studies on ZFSCPN glass system revealed single T_g and single T_c (corresponding to the crystallization of calcium sodium phosphate). Both T_g and T_c decreased when x was increased.

Despite containing Fe_2O_3, CSPFN, MCSPC'F, and ZFSCPN glasses were paramagnetic at room temperature. The reciprocal of the magnetic susceptibility (χ^{-1}) of all glasses as a function of temperature showed the typical Curie–Weiss law behavior of paramagnetic materials [44, 46, 62]. A negative θ_p obtained for all the three series of glasses indicates super-exchange type magnetic interactions between the iron ions in the glassy network. The molar fraction of Fe^{3+} and Fe^{2+} ions in the glasses using Eqs. 1.2–1.4

shows a similar trend in all the three glasses and is a function of x: the concentration of Fe^{3+} initially increased to a maximum value and then continuously decreased with increase in iron content, whereas the concentration of Fe^{2+} ions increased continuously with increasing iron oxide. These data indicate the dominance of superexchange interactions between these ions with increasing iron oxide content in all the glasses [38, 44, 46, 47, 62]. Figure 1.11 shows the variation of Fe^{3+} and Fe^{2+} ion fractions with iron oxide content in the three glass systems.

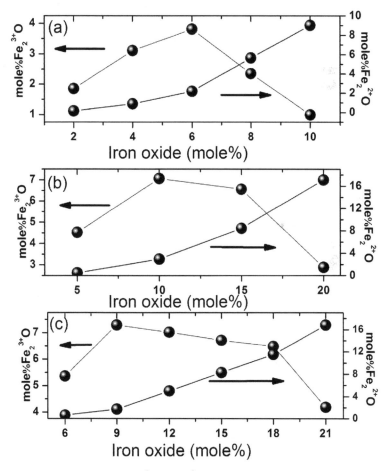

Figure 1.11 Variation of Fe^{3+} and Fe^{2+} ion fraction with iron oxide content (a) CSPFN, (b) MCSPC'F, and (c) ZFSCPN glasses.

Iron ions exist in different valence states with different local symmetries in the glass matrices, for example, as Fe^{3+} with both tetrahedral and octahedral coordination, and/or as Fe^{2+} with octahedral coordination. Fe^{3+} and Fe^{2+} are paramagnetic ions. However, only Fe^{3+} ($3d^5$, $^6S_{5/2}$) shows electron spin resonance (ESR) absorption at room temperature. The EPR absorption spectra of Fe^{3+} ions in oxide glasses are generally characterized by the appearance of resonance absorption at $g \approx 2.1$, 4.3, and 6.0, with their relative intensities being strongly dependent on composition [38, 44, 47]. The $g \approx 2.1$ resonance is assigned to those ions that interact by a super-exchange coupling and can be considered distributed in clusters. The $g \approx 4.3$ resonance line is characteristic of isolated Fe^{3+} ions predominantly situated in rhombically distorted octahedral or tetrahedral oxygen environments [63]. The $g \approx 6.0$ resonance line arises from axially distorted sites.

The main feature in EPR spectra in all silica-based paramagnetic bioactive glasses containing iron oxide, such as CSPFN, MCSPCF, and ZFSCPN glasses, is the presence of resonance at $g \approx 4.3$ and $g \approx 2.1$ (Fig. 1.5). The EPR parameters, i.e., line intensity (J) and linewidth (ΔH), of each absorption line provide information of the associated local environment changes around the Fe^{3+} ion as a function of composition, temperature, etc. Figure 1.12 shows the variations of the EPR parameters with Fe_2O_3 content in CSPFN glasses [44]. We know that the $g \approx 2.1$ resonance arises due to the formation of iron clusters, which give rise to super-exchange type interaction between iron ions. Super-exchange mechanisms tend to narrow the absorption line. On the other hand, interactions between Fe^{3+} and Fe^{2+} ions tend to broaden the linewidth. The final linewidth depends on the relative strengths of the two mechanisms influencing the linewidth. The increase in the linewidth of the $g \approx 2.1$ resonance with Fe_2O_3 concentrations shows the dominance of the broadening mechanisms, which in turns indicates a larger increase in Fe^{2+} ion concentration as Fe_2O_3 concentration is increased in the glass. On the other hand, the slow increase in $g \approx 4.3$ line intensities for low Fe_2O_3 concentrations indicates that fewer Fe^{3+} ions reside in low symmetry sites at these compositions. The decrease in $g \approx 4.3$ line intensity beyond $x = 6$ mole% reflects a reduction in the number of Fe^{3+} ions in octahedral (tetrahedral) environment in the vitreous matrix. In this manner, the subtle changes in the local glassy environment can be inferred from the composition dependence of the EPR parameters of various bioglasses containing iron ions.

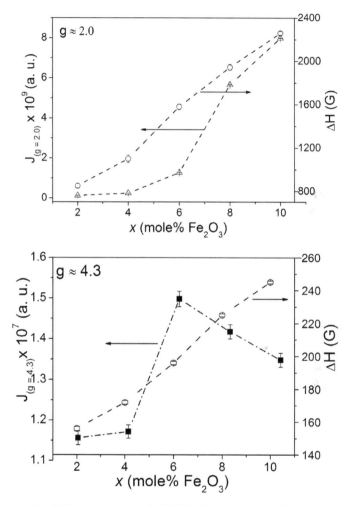

Figure 1.12 EPR parameters of CSPFN glasses. Reprinted from Ref. 44, Copyright 2008, with permission from Elsevier.

The addition of Fe_2O_3 in CaO-SiO_2 glass or an increase in Fe^{3+} ion concentration in Na_2O-CaO-P_2O_5-SiO_2 glass [14, 16] reduces the apatite-forming ability of the respective glasses. On the other hand, the addition of Na_2O or P_2O_5 facilitates apatite formation in CaO-SiO_2-based glasses. It has been reported [3–5, 54, 64] that when iron oxide replaces SiO$_2$ in CSPFN glasses leaving the amount of CaO and P$_2$O$_5$ undisturbed, bioactivity of such systems improved. Such behavior has also been found in MCSPC'F and ZFSCPN glasses [65, 66]. Such studies provide means of increasing the iron content without sacrificing the bioactivity of the glasses. HA formation gradually improves from small-sized crystalline aggregates to a well-crystallized HA phase confirmed by XRD patterns. Monitoring the pH variation during the SBF treatment of glasses provides information on the ion exchange reactions occurring since the time of immersion in SBF. An increase in pH indicates Ca and P ion release from SBF. The pH variation in experiments involving melt-quenched glasses is comparatively slower than that in sol–gel-derived glasses [23, 24]. This is attributed to the high surface reactivity and porosity of sol–gel-derived systems as compared to melt-derived ones. Analysis of the bioactivity tests conducted in several glasses, such as CSPFN, MCSPC'F, and ZFSCPN glasses, shows that 7 days of soaking in SBF is sufficient to yield a well-crystallized HA surface layer. The appearance of the HA peaks and their progressive narrowing indicates this. Similarly, the appearance of the FT-IR band at 970 cm^{-1}, which corresponds to the v_1 P-O symmetric stretching frequency, indicates the obviation of phosphate ions from the ideal tetrahedral structure. This is considered the signature of the formation of calcium phosphate (hydroxyapatite) surface layer. The analysis of the intensity and the area under the bands corresponding to the carbonate and phosphate groups provides information on the evolution and chemical composition of the surface layer.

1.4.2 Magnetic Bioglass-Ceramics

Magnetic bioglass-ceramics are usually derived from the parent bioglass compositions by carefully regulated heat treatment schedule. As a result, nucleation and growth of various crystalline phases occur in the glassy matrix. Hench reported several bioactive glass-ceramics such as A/W glass-ceramics, ceravital glass-ceramics, and machineable glass-ceramics [3–5]. Structural analysis of these glass-ceramics shows that various biomineral phases precipitated along with magnetic phase(s) giving rise to magnetic properties as well as enhanced bioactivity [9–16, 37]. The mineral and magnetic phases that crystallize in various bioglasses depend on the constitution and composition of the glasses. In most of the SiO_2-based quaternary bioglass-ceramics containing Fe_2O_3 (such as those derived from CSPFN glasses discussed earlier), hydroxyapatite, wollastonite, and magnetite are the crystalline mineral phases that precipitate during the heat treatment. In sol–gel-derived glass-ceramics, some intermediate (bioactive mineral) phases such as sodium calcium silicate and calcium phosphate have been reported [67, 68]. Hydroxyapatite, wollastonite, akermanite, and magnetite crystallize when MCSPC′F glasses are subject to heat treatment [45]. Zinc ferrite [64, 66] and cristobalite [22, 68] have been observed when Zn and Fe, and Na containing glasses are heat treated, respectively. The heat treatment temperature (T_A) and time of heat treatment (t_A) play an important role in the crystallization process of glass. Usually crystallization temperature is chosen from the thermal analysis of the parent glass. t_A up to several hours at T_A corresponding to the T_c of various crystalline phases yields nanocrystals whose average crystallite increases with t_A and T_A. In MgBGCs derived from CSPFN glass by heat treatment at 1050°C for 3 h, the average crystallite size of the magnetite phase varied from 32 to 56 nm as a function of iron oxide content. Since the bone mineral (HA and W) content also increases with increase in t_A and T_A, such treatment improves the bioactivity of the glass-ceramics. The presence of magnetite nanocrystallites (which are formed by the reduction of Fe_2O_3 when the glass is heat treated in oxygen-deficient environment inside an electric furnace) induces magnetism in these glass-ceramics and hence their name MgBGC. Magnetic properties such as saturation magnetization (M_s), coercivity (H_c), hysteresis area loop, iron ion fraction, and iron ions interactions are important to evaluate these for application as thermoseeds. M_s generally increases when iron content increases because the magnetic phase present in the glassy system increases. MgBGCs derived from CSPFN glass containing 10 mole% iron oxide exhibit a saturation magnetization (M_s) of 7.95 emu/g for an applied magnetic field of 20 kOe [69]. In MgBGCs derived from MCSPC′F glass containing 20 mole% iron oxide, M_s of 8.89 emu/g has been reported [70] at 20 kOe, whereas in ZFSCPN MgBGC, a maximum M_s of 19.60 emu/g has been observed for 21 mole% of iron content at 20 kOe [71]. An increase in the hysteresis loop area with iron content has been observed in all MgBGCs, whereas coercivity showed a decrement, which indicates the soft magnetic nature on these ceramics. 25SiO$_2$50CaO15P$_2$O$_5$(10-x)Fe$_2$O$_3$$x$ZnO glass-ceramics with up to 5 mole% ZnO showed a transition from ferromagnetic to ferrimagnetic behaviour with increase in ZnO content [71]. Heat treatment of the constituents at 800°C for 6 h produced spherical and evenly dispersed nanometer-sized calcium phosphate, hematite, and magnetite crystallites in the glassy silica matrix [71]. Coercivity of nanocrystalline magnetic oxide dispersed in a glassy matrix does not follow the universal behavior of nanocrystalline magnetic oxides as proposed by Maaz et al. [72]. A temperature-dependent phase conversion of ferrimagnetic Fe$_3$O$_4$ into nonmagnetic hematite leading to irregularity in the coercive field and decrease in M_s has been reported in CSPFN MgBGCs [38]. A magnetic susceptibility study of sol–gel-derived aluminosilicate glass-ceramics revealed antiferromagnetism and an increased conversion of Fe^{2+} ions into Fe^{3+} ions with increasing T_A [73].

EPR and Mossbauer spectroscopy have been employed to understand the distribution of Fe^{3+} and Fe^{2+} ions in various glass matrices. The EPR spectra of CSPFN glass heat treated at different T_A show the nature of Fe in distribution and the interaction between the Fe ions. A decrease in J and ΔH of the absorption at $g \approx 4.3$ and their ultimate disappearance above T_A = 850°C have been attributed to the dominance of clustering mechanism as the crystallization of magnetite content increases. A significant increase in J and ΔH of the $g \approx 2.1$ absorption line supports this argument. For T_A > 1050°C, the decrease in J and ΔH of the $g \approx 2.1$ absorption line taken together with the decrease in M_s and H_c (despite a continuous increase in average crystallite size) indicate the precipitation of a nonmagnetite phase in the glassy matrix [38, 47]. The presence of such nonmagnetic ions suppresses the super-exchange interaction of the neighbors leading to the observed decrease in the EPR parameters [38, 47]. The same behavior has been observed in glass containing zinc–iron ferrite when heat treated under different environments [74].

Glass-ceramics exhibit better bioactivity as compared to their parent glasses because of the presence of bone mineral phases [3–5]. Both *in vitro* and *in vivo* have been performed on glass-ceramics to confirm their better applicability as implant materials and thermoseeds. The assumption that the addition of metal ions to ferrimagnetic bioglass-ceramics accelerates the apatite formation in SBF has been confirmed [75]. MgBGCs derived by heat treatment of

CSPFN, MCSPC'F, and ZFSCPN glasses exhibited excellent bioactivity as well as ferromagnetic properties [37]. Figure 1.13 shows the results of *in vitro* bioactivity test performed on one composition of MgBGCs derived from CSPFN glass [76]. A systematic monitoring of pH variation during *in vitro* tests has been performed for the understanding of surface kinetic reactions in these systems. It has been observed that in the early stage of immersion (2 days), the concentrations of Ca, P, and Si ions increased in the SBF solution in a specific manner. These results indicate the release of alkali and alkaline earth ions, loss of soluble SiO_2 from the surface of the glass specimens to the SBF solution, condensation, and repolymerization of SiO_2-rich layer [77]. After 7 days of immersion, Ca and P concentrations in the SBF solution decreased, whereas Si ion concentration increased. This shows the formation of the amorphous $CaO-P_2O_5$ layer on the surface of the glass sample. Subsequent rapid decrease in P and Ca ion concentrations in SBF solution after 14 days immersion indicated crystallization and growth of the $CaO-P_2O_5$ rich layer. These stages occur faster for systems with higher bioactivity [76]. Similarly, the addition of ZnO to the basic composition is also assumed to accelerate the formation of apatite on glass-ceramics. Zinc ions migrate into the fluid as $Zn(OH)_2$ and contribute more OH^- ions required for the crystallization of the amorphous Ca-P from SBF to form the apatite layer [66].

A recent study confirmed the acceleration of the formation of the bioactive surface layer on ferrimagnetic glass-ceramic with a basic composition of $40Fe_2O_3 15P_2O_5 20SiO_2 5TiO_2$ through the addition of 20% of different metal oxides like MgO, CaO, MnO, CuO, ZnO, or CeO_2 [77]. A quantitative analysis of the ions in the solution after *in vitro* tests reveals that the concentrations of Ca and P elements in SBF solutions with glass samples decreased or increased with the ability of the glass samples for forming HA layer.

In addition to the response of MgBGCs in SBF, it is also important to study their response in biopolymers. Serum albumin, which is the most abundant protein found in human blood, becomes the natural choice for studying the response. The absorption of protein in bioglass/glass-ceramics usually takes place by the physical adsorption process. Thus, interactions are mainly influenced by the surface properties of the materials. These interactions can be demonstrated by different surface analytical techniques like ESCA, SEM, SIMS, etc.

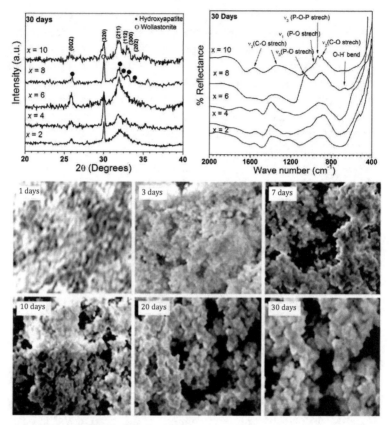

Figure 1.13 GI-XRD patterns, FT-IR spectra, and SEM micrographs of MgBGCs derived from CSPFN glasses after treatment in SBF for 30 days.

SEM micrographs of the MCSPC'F glass-ceramics samples after immersion in BSA for 7 days are shown in Fig. 1.14a–c. The surface gets covered with polymeric layer (BSA) as iron oxide content is increased from 10 to 15 wt.%. Some interconnecting pores of 100–500 nm were observed, implying that the surface was not uniformly covered with polymer. SIMS spectra of glass-ceramics samples after immersion in BSA for 7 days are shown in Fig. 1.14d. The MCSPC'F10 surface showed Ca (m/z = 40) and Fe (m/z = 56) as major peaks, while some peaks corresponding to C and Si were also visible. In addition, peaks corresponding to polymeric chains were observed, when iron oxide content was increased to 15–20 wt%. The peak around (m/z = 69) was assigned to Ga ions (ion source), which were used for bombardment of the samples. Secondary positive ion peaks corresponding to CH_3, CHN, CO, $CONH_2$, $C_2H_5NH_2$, CN_3H_3, C_2H_6NO, $C_2H_2CONH_2$, $C_3H_4CONH_2$, related to adsorbed BSA, are evident at m/z = 15, 27, 28, 44, 45, 57, 60, 70, and 84, respectively. These fragments (CH2–CH2–NH2 and others) related to the adsorption of BSA were not observed on the MCSPC'F10 surface. These materials thus exhibited the property that can be exploited to bind specific proteins on the surface for magnetic drug targeting or polymer detection and separation.

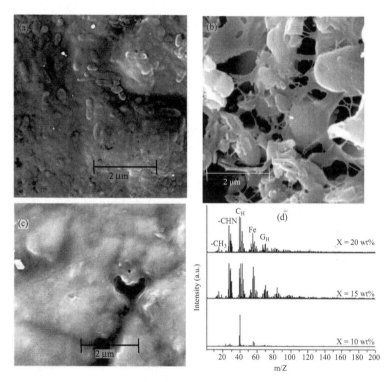

Figure 1.14 SEM of glass-ceramics samples (a) MCSPC'F10, (b) MgBGCF15, and (c) MCSPC'F20; (d) SIMS of MCSPC'F after 7 days immersion in BSA. Reprinted from Ref. 51, Copyright 2012, with permission from Elsevier.

To impart multiple functionalities, magnetic particles are coated with noble metals, for example silver or gold, for enhancing the optical properties due to plasmon resonance effects [78]. In addition, silver is known to control the bacterial growth in a variety of applications, including dental work, catheters, and burn wounds. Ag ions are toxic to micro-organisms, exhibiting strong biocidal effects in as many as 12 species of bacteria, including *Escherichia coli* [79]. Glass/glass-ceramics can be potential carriers of antibacterial ions like silver because these materials can accommodate these ions in their structure. The rate of release of Ag ions can be controlled by modifying the structure of the materials or using some external stimuli like magnetic or electric field. Such a procedure has been demonstrated by studying the elemental release from glass-ceramics samples in DI water [80]. Elemental release from different glass-ceramics samples after 1 day immersion in DI water is given in Table 1.5. It was observed that silica, PO_4^{3-}, Fe, and Ag leach out from the surface. Increase in concentrations of Fe and Ag after immersion was more on ZFSCPAg4 as compared to ZFSCPAg2. The increase in Ag release is expected as the total amount of Ag content in the matrix was more.

Table 1.5 Ion release from ZFSCPAg after 1 day immersion in DI water

Sample	Si (µg/ml)	P (µg/ml)	Fe (µg/ml)	Ag (µg/l)
ZFCPAg2	13 ± 0.03	12 ± 0.06	7 ± 0.01	25 ± 0.41
ZFCPAg4	9 ± 0.02	10 ± 0.05	12 ± 0.02	37 ± 2.63

The antibacterial activity of Ag-containing glass-ceramics evaluated using *E. coli* MG1655, Gram-negative coliform bacteria, is shown in Table 1.6. The released Ag ions from the ceramic attach to the negatively charged bacterial cell wall and rupture it, leading to cell death [81]. It has also been hypothesized that oxygen associates with silver and reacts with the sulfhydryl (–S–H) groups on cell wall to form R–S–S–R bonds, subsequently blocking respiration and causing cell death [82]. The glass-ceramic containing 4% Ag showed complete inhibition of bacterial growth at a concentration of 10% (w/v); however, the glass-ceramic containing 2% Ag showed the similar effect at 20% (w/v) (Table 1.6).

The assembly of individual magnetic particles takes place under dipole–dipole interactions or when subjected to external magnetic field, which can lead to net higher magnetization [83, 84]. It has been recently reported that the magnetic nanoparticles form magnetic assembly when decorated with the gold particles [85, 86]. The microstructures of glass-ceramics containing different amount of Au content are shown in Fig. 1.15. Microstructures of glass-ceramics samples ZFSCPAu1 and ZFSCPAu2 show the formation of small-sized particles, which are homogeneously distributed (see Fig. 1.15). However, the microstructure of ZFSCPAu4 revealed the formation of assembly of magnetic particles.

Table 1.6   Antibacterial tests of glass-ceramics (ZFSCPAg) against *E. coli*

Concentration (%w/v) of glass in LB broth	Colony-forming unit (cfu/ml) in presence of ZFCPAg2	Colony-forming unit (cfu/ml) in presence of ZFCPAg4
1	7.4×10^8	3.7×10^8
5	2.1×10^6	1.6×10^5
10	4.5×10^3	Nil
15	2.3×10^2	Nil
20	Nil	Nil

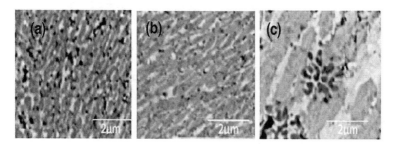

Figure 1.15 Microstructure of samples (a) ZFSCPAu1, (b) ZFSCPAu2, and (c) ZFSCPAu4. Reprinted from Ref. 86, Copyright 2013, with permission from Elsevier.

The development of ferrimagnetic crystallites within a glass phase is controlled by the nucleation and growth of the magnetic particles. The addition of Au in glass matrix induces nucleation and thus can influence the Fe distribution at different crystallographic sites. Magnetic properties in a spinel-like phases depend on the arrangement of Fe^{2+} and Fe^{3+} ions in the spinel lattice [86]. This will affect the interparticle interactions among the particles. The magnetization values of glass-ceramics samples measured at room temperature are shown in Fig. 1.16. Magnetization and coercive field values increased with the increase of Au content. This increase is attributed to the increase in the formation of magnetite assembly of particles. Due to the enhanced bone-forming bioactivity, degradation and drug-delivery capability, coupled with magnetic properties, MgBGCs find potential application as thermoseeds in hyperthermia application and as local drug-delivery carriers. Tumor hyperthermia requires accurate calculation and description of the temperature distribution induced by alternating magnetic field in the tissue before it is applied in clinical trial.

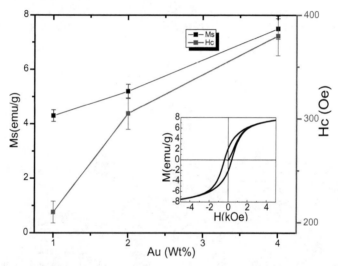

Figure 1.16 Variation of M_s and H_c with gold content; representative M-H plot for ZFSCPAu4 sample is shown in the inset.

In a study of zinc ferrite MgBGCs, a maximum specific power loss of 31.5 W/g corresponding a temperature rise to 45°C after exposure to an alternating field of 500 Oe for 120 s has been reported [87]. Antimicrobial property has been evaluated in MgBGCs containing Ag ions [80] through *E. coli.* cell culture. The sustained drug-delivery capacity of MgBGCs suitable for their *in vitro* application in local drug therapy has also been demonstrated [88]. pH variation during *in vitro* testing of iron oxide nanoparticles coated with silica prepared through sol–gel route points toward their application as contrast agent in MRI, magnetic drug targeting, and hyperthermia anticancer strategy [89]. Attempts have been made to prepare magnetic nanoparticles coated with bioactive materials by gas flame, micro-emulsion, sol–gel, and laser spinning techniques and evaluate them for biomedical applications [18–23, 25–27].

Summary

We have reviewed the developments in bioactive glasses and glass-ceramics containing iron oxide. The influences of elemental constitution, composition variation, and processing conditions on the properties of these biomaterials have been outlined bearing the applications in mind. Although osteo application is fundamental to biomaterials, hyperthermia, drug delivery, and MRI applications are specific to magnetic biomaterials.

Considerable progress has been made in the last 10 years to develop MgBGCs for biomedical applications. However, our current understanding of the long-term *in vitro* and *in vivo* characteristics of these systems, specifically the long-term effect of degradation and ion release kinetics of inorganic bioactive phases, is still limited. The fixation of orthopedic implants has been the most difficult and challenging problem. However, the promise shown by these MgBGCs motivates continued focus on developing bioglasses and MgBGC with enhanced magnetic and bioactive properties. Glasses with more than 60% silica content, capable of accommodating higher magnetic content, would be one of the points of focus of materials scientists. Since preparative methods have an important role in determining the magnetic and bioactive properties, attempts to prepare MgBGCs through less explored techniques, such as flame, micro-emulsion, sol–gel, and laser spinning techniques, could yield MgBGCs more suited for biomedical applications. The development of appropriate characterization techniques, coupled with predictive analytical models, is another topic that requires immediate attention. A combinatory approach, possibly using stem cells, signaling molecules, and novel functional biomaterials, needs to be adapted into tissue engineering. Although bioactive glasses are mechanically weak, it has recently been discovered that 45S5 Bioglass can partially crystallize when heated to high temperatures (>950°C) during scaffold fabrication and that the mechanically strong crystalline phase can transform to a biodegradable, amorphous calcium phosphate at body temperature and in a biological environment [90]. This transformation enables the two normally irreconcilable properties—mechanical competence and biodegradability—to be combined in a single scaffold. Hence, engineered composite scaffolds made by the smart combination of biodegradable polymers and bioactive glasses need to be explored. Besides possible toxicity issues associated with nanoparticles, synthetic polymers and carbon nanotubes may also improve the environment to enhance cell attachment and proliferation as well as adding extra functionalities to the base scaffold. These approaches promise to us toward scaffold optimization and its clinical application.

Acknowledgments

One of the authors (AS) acknowledges financial support from the Board of Research in Nuclear Sciences, Department of Atomic Energy, India, through collaborative research projects (Nos. 2005/34/15/BRNS and 2010/34/54/BRNS) and the Department of Science and Technology, India, vide project No: SR/S2/CMP-19/2006.

References

1. Shelby, J. E. (1997). *Introduction to Glass Science and Technology* (The Royal Society of Chemistry, Cambridge, UK).
2. Lewis, M. H. (1988). *Glass and Glass Ceramics* (Champon and Hall, London).
3. Hench, L. L. (1991). Bioceramics: From concept to clinic. *J. Am. Ceram. Soc.*, **74**, pp. 1487–1510.
4. Hench, L. L. (1998). Bioceramics. *J. Am. Ceram. Soc.*, **81**, pp. 1705–1728.
5. Hench, L. L. (1998). Biomaterials: A forecast for the future. *Biomaterials*, **19**, pp. 1419–1423.
6. Aza, P. N., Guitián, F., and Aza S. (1997). Bioeutectic: A new ceramic material for human bone replacement. *Biomaterials*, **18**, pp. 1285–1291.
7. Day, R. M. (2005). Bioactive glass stimulates the secretion of angiogenic growth factors and angiogenesis *in vitro*. *Tissue Eng.*, **11**, pp. 768–777.

8. Rezwan, K., Chen, Q. Z., Blaker, J. J., and Boccaccini, A. R. (2006). Biodegradable and bioactive porous polymer/inorganic composite scaffolds for bone tissue engineering. *Biomaterials*, **27**, pp. 3413–3431.
9. Ohura, K., Ikenaga, M., Nakamura, T., Yamamuro, T., Ebisawa, Y., Kokubo, T., Kotoura, Y., and Oka, M. (1991). A heat-generating bioactive glass-ceramic for hyperthermia. *J. Appl. Biomater.*, **2**, pp. 153–159.
10. Ebisawa, Y., Miyaji, F., Kokubo, T., Ohura, K., and Nakamura, T. (1997). Bioactivity of ferrimagnetic glass-ceramics in the system $FeO-Fe_2O_3$-$CaO-SiO_2$. *Biomaterials*, **18**, pp. 1277–1284.
11. Arcos, D., Real, R. P., and Vallet, R. M. (2002). A novel bioactive and magnetic biphasic material. *Biomaterials*, **23**, pp. 2151–2158.
12. Leventouri, T., Kis, A. C., Thompson, J. R., and Anderson, I. M. (2005). Structure, microstructure, and magnetism in ferrimagnetic bioceramics. *Biomaterials*, **26**, pp. 4924–4931.
13. Chun, H. H., Ching, W. C., Sheng, M. H., Yu, T. L., and Feng, H. L. (2009). The fabrication and characterization of dicalcium phosphate dihydrate-modified magnetic nanoparticles and their performance in hyperthermia processes *in vitro*. *Biomaterials*, **30**, pp. 4700–4707.
14. Ebisawa, Y., Sugimoto, Y., Hayashi, T., Kokubo, T., Ohura, K., and Yamamuro, T. (1991). Crystallization of (FeO, Fe_2O_3)-$CaO-SiO_2$ glasses and magnetic properties of their crystallized products. *J. Ceram. Soc. Jpn.*, **99**, pp. 7–13.
15. Ebisawa, Y., Miyaji, F., Kokubo, T., Ohura, K., and Nakamura, T. (1997). Surface reactivity of bioactive and ferrimagnetic glass-ceramics in the system $FeO-Fe_2O_3$-$CaO-SiO_2$. *J. Ceram. Soc. Jpn.*, **105**, pp. 947–951.
16. Kawashita, M., Tanaka, M., Kokubo, T., Inoue, Y., Yao, T., Hamada, S., and Shinjo, T. (2005). Preparation of ferrimagnetic magnetite microspheres for *in situ* hyperthermic treatment of cancer. *Biomaterials*, **26**, pp. 2231–2238.
17. Jones, R. W. (1989). *Fundamental Principles of Sol–Gel Technology* (The Institute of Metals, London).
18. Oki, A., Parveen, B., Hossain, S., Adeniji, S., and Donahue, H. (2004). Preparation and *in vitro* bioactivity of zinc containing sol–gel-derived bioglass materials. *J. Biomed. Mater. Res. A*, **69**, pp. 216–221.
19. Fooladi, A. A., Hosseini, H. M., Hafezi, F., Hosseinnejad, F., and Nourani, M. R. (2013). Sol-gel-derived bioactive glass containing SiO_2-MgO-CaO-P_2O_5 as an antibacterial scaffold. *J. Biomed. Mater. Res. A*, **101**, pp. 1582–1587.

20. Rahpeyma, S., Fathi, M. H., Ebrhimi, K. R., and Doustmohannadi, A. (2013). Fabrication and characterization of sol–gel derived bioactive glass/zirconia composite coating for biocompatibility improvement of AISI 316L stainless steel. *J. New Mater.*, **3**, pp. 93–104.
21. Lenza, F. S., and Vasconcelos, L. W., (2002). Synthesis of titania-silica materials by sol–gel. *Mater. Res.*, **5**, pp. 497–502.
22. Yang, X., Zhang, L., Chen, X., Sun, X., Yang, G., Guo, X., Yang, H., Gao, C., and Gou, Z. (2012). Incorporation of B_2O_3 in CaO-SiO_2-P_2O_5 bioactive glass system for improving strength of low-temperature co-fired porous glass ceramics. *J. Non-Cryst. Solids*, **358**, pp. 1171–1179.
23. Nezafati, N., Moztarzadeh, F., and Hesaraki, S. (2012). Surface reactivity and *in vitro* biological evaluation of sol–gel derived silver/calcium silicophosphate bioactive glass. *Biotechnol. Bioprocess Eng.*, **17**, pp. 746–754.
24. Shankhwar, N., Kothiyal, G. P., and Srinivasan, A. (2015). Influence of phosphate precursors on the structure, crystallization behaviour and bioactivity of sol–gel derived 45S5 Bioglass. *RSC Advances*, **5**, pp. 100762–100768.
25. Zhu, M., Shi, J., He, Q., Zhang, L., Chen, F., and Chen, Y. (2012). An emulsification-solvent evaporation route to mesoporous bioactive glass microspheres for bisphosphonate drug delivery. *J. Mater. Sci.*, **47**, pp. 2256–2263.
26. Malik, M. A., Wani, M. Y., and Hashim, M. A. (2012). Microemulsion method: A novel route to synthesize organic and inorganic nanomaterials. *Arabian J. Chem.*, **5**, pp. 397–417.
27. Quintero, F., Pou, J., Comesaña, R., LusquiñosAz, F., Riveiro, A., Mann, A. B., Hill, G. H., Wu, Z. Y., and Jones, J. R. (2009). Laser spinning of bioactive glass nanofibers. *Adv. Funct. Mater.*, **19**, pp. 3084–3090.
28. Chen, Q. Z., and Boccaccini, A. R. (2008). Tissue engineering scaffolds from bioactive glass and composite materials, in *Topics in Tissue Engineering* (Ashammakhi, N., Reis, R., and Chiellini, F.), vol 4.
29. Abdollahi, S., Ma, A. C., and Cerruti, M. (2013). Surface transformations of Bioglass 45S5 during scaffold synthesis for bone tissue engineering. *Langmuir*, **29**, pp. 1466–1474.
30. Shih, S.-J., Chou, Y.-J., Chen, C.-Y., and Lin, C.-K. (2013). One-step synthesis and characterization of nanosized bioactive glass. *J. Med. Biol. Eng.*, **34**, pp. 18–23.
31. Suchanek, W. L., and Riman, R. E. (2006). Hydrothermal synthesis of advanced ceramic powders. *Adv. Sci. Technol.*, **45**, pp. 184–193.

32. Li, K., and Tjong, S. C. (2011). Hydrothermal synthesis and biocompatibility of hydroxyapatite nanorods. *J. Nanosci. Nanotechnol.*, **11**, pp. 10444–10448.
33. Cullity, B. D. (1978). *Elements of X-Ray Diffraction*, 2nd ed. (Addison-Wesley Pub., Reading).
34. Reitvild, H. M. (1969). A profile refinement method for nuclear and magnetic structures. *J. Appl. Ceys.*, **2**, pp. 65–71.
35. Klug, H. P., and Alexander, L. E. (1974). *X-ray Diffraction Procedures for Polycrystalline and Amorphous Materials*, 2nd ed. (John Wiley & Sons Inc.).
36. Williamson, G. K., and Hall, W. H. (1953). X-ray line broadening from filed aluminium and wolfram. *Acta Metall.*, **1**, pp. 22–31.
37. Singh, R. K. (2009). Investigations on selected bioactive glasses and glass ceramics containing iron oxide, Ph.D. thesis, Indian Institute of Technology Guwahati, India.
38. Shankhwar, N., Singh, R. K., Kothiyal, G. P., Perumal, A., Srinivasan, A. (2014). Evolution of magnetic properties of $CaO-P_2O_5-Na_2O-Fe_2O_3-SiO_2$ glass upon heat treatment. *IEEE Trans. Magn.*, **50**, 4003504.
39. Rodriguez, J. C. (1993). Recent advances in magnetic structure determination by neutron powder diffraction, *Physica B*, **192**, pp. 55–69.
40. Wu, J. (2013). Application of electron diffraction in the structure characterization of polymer crystals. *Chinese J. Polym. Sci.*, **31**, pp. 841–852.
41. Coats, A. W., and Redfern, J. P. (1963). Thermogravimetric analysis. A review. *The Analyst*, **88**, pp. 906–924.
42. http://www.netzsch-thermal-analysis.com/en/landing-pages/differential-scanning-calorimetry-dsc-differential-thermal-analysis-dta.html.
43. Riga, A. T., and Neag, C. M. (1991). *Materials Characterization by Thermomechanical Analysis* (ASTM International, Philadelphia).
44. Singh, R. K., Kothiyal, G. P., and Srinivasan, A. (2008). Electron spin resonance and magnetic studies on $CaO-SiO_2-P_2O_5-Na_2O-Fe_2O_3$ glasses, *J. Non-Cryst. Solids*, **354**, pp. 3166–3170.
45. Singh, R. K., and Srinivasan, A. (2009). EPR and magnetic properties of $MgO-CaO-SiO_2-P_2O_5-CaF_2-Fe_2O_3$ glass-ceramics. *J. Magn. Magn. Mater.*, **321**, pp. 2749–2752.

46. Singh, R. K., and Srinivasan, A. (2010). EPR and magnetic susceptibility studies of iron ions in ZnO-Fe$_2$O$_3$-SiO$_2$-CaO-P$_2$O$_5$-Na$_2$O glasses. *J. Magn. Magn. Mater.*, **322**, pp. 2018–2022.

47. Shankhwar, N., Kothiyal, G. P., and Srinivasan, A. (2014). Understanding the magnetic behaviour of heat treated CaO-P$_2$O$_5$-Na$_2$O-Fe$_2$O$_3$-SiO$_2$ bioactive glass using electron paramagnetic resonance studies. *Physica B*, **448**, pp. 132–135.

48. Cullity, R. D. (1972). *Introduction to Magnetic Materials* (Addison-Wesley, Reading).

49. Watts, J. F., and Wolstenholme, J. (2003). *An Introduction to Surface Analysis by XPS and AES* (Wiley & Sons, Chichester, UK. Benninghoven).

50. Benninghoven, A., Rüdenauer, F. G., and Werner, H. W. (1987). *Secondary Ion Mass Spectrometry: Basic Concepts, Instrumental Aspects, Applications, and Trends* (Wiley, New York).

51. Sharma, K., Bhattacharya, S., Murali, C., and Kothiyal, G. P. (2012). Interactions of ferrimagnetic glass/glass-ceramics with bovine serum albumin. *Appl. Surf. Sci.*, **258**, pp 2356–2361.

52. Ohtsuki, C., Kokubo, T., and T. Yamamuro. (1992). Mechanism of apatite formation on CaO-SiO$_2$-P$_2$O$_5$ glasses in a simulated body fluid. *J. Non-Cryst. Solids*, **143**, pp. 84–92.

53. Kokubo, T., and Hiroaki, T. (2006). How useful is SBF in predicting *in vivo* bone bioactivity? *Biomaterials*, **27**, pp. 2907–2915.

54. Singh, R. K., Kothiyal, G. P., and Srinivasan, A. (2009). *In vitro* evaluation of bioactivity of CaO-SiO$_2$-P$_2$O$_5$-Na$_2$O-Fe$_2$O$_3$ glasses. *Appl. Surf. Sci.* **255**, pp. 6827–6831.

55. Sharma, K., Deo, M. N., and Kothiyal, G. P. (2012). Effect of iron oxide addition on structural properties of calcium silico phosphate glass/glass-ceramics. *J. Non-Cryst. Solids*, **358**, pp. 1886–1891.

56. Sharma, K., Dixit, A., Singh, S., Prajapat, C. L., Bhattacharya, S., Jagannath, Singh, M. R., Yusuf, S. M., Sharma, P. K., Tyagi, A. K., and Kothiyal, G. P. (2009). Preparation and studies on surface modifications of calcium-silico-phosphate ferrimagnetic glass-ceramics in simulated body fluid. *J. Mater. Sci. Eng. C*, **29**, pp. 2226–2233.

57. Park, J., and Lakes, R. S. *Biomaterials*, 3rd ed. (Springer, USA).

58. Park, J. (2008). *Bioceramics: Properties, Characterizations and applications* (Springer, USA).

59. Ikenaga, M., Ohura, K., Yamamuro, T., Kotoura, Y., Oka, M., and Kokubo, T. (1993). Localized hyperthermic treatment of experimental bone tumors with ferromagnetic ceramics. *J. Orthop. Res.*, **11**, pp. 849–855.

60. Kokubo, T. (1991). Bioactive glass ceramics: Properties and applications. *Biomaterials,* **12**, pp. 155–163.
61. Hench, L. L. (1998). Bioceramics: Materials characteristics versus in-vivo behavior. *Ann. New York Acad. Sci.*, **523**, pp. 54.
62. Singh, R. K., and Srinivasan, A. (2010). Structural and magnetic properties of MgO-CaO-SiO$_2$-P$_2$O$_5$-CaF$_2$-Fe$_2$O$_3$ glasses. *Eur. J. Glass Sci. Technol.,* **51**, pp. 778–782.
63. Castner, T., Newell, G. S., Holtan W. C., and Slichter, C. P. (2004). Note on the paramagnetic resonance of iron in glass. *J. Chem. Phys.* **32**, pp. 668–673.
64. Salwa, A. M., Hameed, A., Abeer, M., and El Kady, E. I. (2012). Effect of different additions on the crystallization behavior and magnetic properties of magnetic glass-ceramic in the system Fe$_2$O$_3$-ZnO-CaO-SiO$_2$. *J. Adv. Res.*, **3**, pp. 167–175.
65. Singh, R. K., and Srinivasan, A. (2011). *In vitro* evaluation of bioactivity of MgO-CaO-SiO$_2$-P$_2$O$_5$-CaF$_2$-Fe$_2$O$_3$ glasses. *Eur. J. Glass Sci. Technol.* Part B, **52**, pp. 47–52.
66. Singh, R. K., and Srinivasan, A. (2009). Bioactivity of SiO$_2$-CaO-P$_2$O$_5$-Na$_2$O glasses containing zinc-iron oxide, *Appl. Surf. Sci.*, **256**, pp. 1725–1730.
67. Chen, Q. Z., and Thouas, G. A. (2011). Fabrication and characterization of sol–gel derived 45S5 Bioglass®-ceramic scaffolds, *Acta Biomaterialia,* **7**, pp. 3616–3126.
68. Cacciotti, I., Mariangela, L., Alessandra, B., Antonio, R., and Laura, M. (2012). Sol–gel derived 45S5 bioglass: Synthesis, microstructural evolution and thermal behavior. *J. Mater. Sci: Mater. Med.*, **23**, pp.1849–1866.
69. Singh, R. K., Kothiyal, G. P., and Srinivasan, A. (2008). Magnetic and structural properties of CaO-SiO$_2$-P$_2$O$_5$-Na$_2$O-Fe$_2$O$_3$ glass ceramics. *J. Magn. Magn. Mater.*, **320,** pp. 1352.
70. Singh, R. K., and Srinivasan, A. (2010). Bioactivity of ferrimagnetic MgO-CaO-SiO$_2$-P$_2$O$_5$-Fe$_2$O$_3$ glass-ceramics. *Ceram. Inter.* **36,** pp. 283–290.
71. Singh, R. K., and Srinivasan, A. (2011). Magnetic properties of bioactive glass-ceramics containing nanocrystalline zinc ferrite. *Magn. Magn. Mater,* **323**, pp. 330–333.
72. Maaz, K., Mumtaz, A., Hasanain, S. K., and Ceylan, A. (2006). Synthesis and magnetic properties of cobalt ferrite (CoFe$_2$O$_4$) nanoparticles

prepared by wet chemical route. *J. Magn. Magn. Mater.*, **308**, pp. 289–295.

73. Coroiu, I., Culea, E., and Darabont, Al. (2005). Magnetic and structural behaviour of the sol–gel-derived iron aluminosilicate glass-ceramics. *J. Magn. Magn. Mater.*, **290**, pp. 997–1000.

74. Kawashita, M., Iwahashi, Y., Kokubo, T., Yao, T., Hamada, S. and Shinjo, T. (2004). Preparation of glass-ceramics containing ferrimagnetic zinc-iron ferrite for the hyperthermal treatment of cancer. *J. Ceram. Soc. Japan*, **112**, pp. 373–379.

75. Ebisawa, Y., Kokubo, T., Ohura, K., and Yamamuro, T. (1990). Bioactivity of CaO SiO$_2$-based glasses: *In vitro* evaluation. *J. Mat. Sci: Mater. Med.*, **1**, pp. 239–244.

76. Singh, R. K., Kothiyal, G. P., and Srinivasan, A. (2008). Evaluation of CaO-SiO$_2$-P$_2$O$_5$-Na$_2$O-Fe$_2$O$_3$ bioglass-ceramics for hyperthermia application. *J. Mat. Sci: Mater. Med.* **20**, pp. 147–151.

77. Marzouk, M. A., and Elwan, R. L. (2014). *In vitro* evaluation of some types of erromagnetic glass ceramics. *Inter. J. Biomater.*, **2014**.

78. Beigi, H. M., Yaghmaei, S., Roostaazad, R., and Arpanae, A. (2013). Comparison of different strategies for the assembly of goldcolloids onto Fe$_3$O$_4$@SiO$_2$ nanocomposite particles. *Physica E*, **49**, pp. 30–38.

79. Hindi, K. M., Ditto, A. J., Panzer, M. J., Medvetz, D. A., Hovis, C. E., Hilliard, J. K., Taylor, J. B., Yun, Y. H., Cannon, C. L., and Youngs, W. L. (2009). The antimicrobial efficacy of sustained release silver–carbene complex-loaded L-tyrosine polyphosphate nanoparticles: Characterization, *in vitro* and *in vivo* studies. *Biomaterials*, **30**, pp. 3771–3779.

80. Sharma, K., Meena, S. S., Saxena S., Yusuf, S. M., Srinivasan A., Kothiyal, G. P. (2012). Structure and magnetic properties of glass-ceramics containing silver and iron oxide, *Mater. Chem. Phys.*, **133**, pp. 144–150.

81. Davies, R. L., and Etris, S. F. (1997). The development and functions of silver in water purification and disease control. *Catalysis Today*, **36**, pp. 107–114.

82. Kumar, V. S., Nagaraja, B. M., Shashikala, V., Padamshri, A. H., Madhavendra, S. S., Raju, B. D., and Rama Rao, K. S. (2004). Highly efficient Ag/C catalyst prepared by electro-chemical deposition method in controlling microorganisms in water. *J. Mol. Cata. A: Chem.*, **223**, pp. 313–319.

83. Ahmed, W., Laarman, R. P. B., Hellenthal, C., Kooij, E. S., Silfhout, A. V., and Poelsema, B. (2010). Dipole directed ring assembly of Ni-coated Au-nanorods. *Chem. Commun.*, **46**, pp. 6711–6713.

84. Lu, Y., Dong, L., Zhang, L. C., Su, Y. D., and Yu, S. H. (2012). Biogenic and biomimetic magnetic nanosized assemblies. *Nano Today*, **7**, pp. 297–315.
85. Lartigue, L., Hugounenq, P., Alloyeau, D., Clarke, S. P., Lévy, M., Bacri, J. C., Bazzi, R., Brougham, D. F., Wilhelm, C., and Gazeau, F. (2012). Cooperative organization in iron oxide multi-core nanoparticles potentiates their efficiency as heating mediators and MRI contrast agents. *ACS. Nano*, **6**, pp. 10935–10949.
86. Sharma, K., Prajapat, C. L., Meena, S. S., Singh, M. R., Yusuf, S. M., Montagne, L. and Kothiyal, G. P. (2013). Influence of Au addition on magnetic properties of iron oxide in a silica-phosphate glass matrix. *J. Magn. Magn. Mater.*, **345**, pp. 24–28.
87. Shah, A. S., Hashmi, M. U., Shamim, A., and Alam, S. (2010). Study of an anisotropic ferrimagnetic bioactive glass ceramic for cancer treatment, *Appl. Phys. A*, **100**, pp. 273–280.
88. Zhu, M., Zhang, J., Zhou, Y., Liu, Y., He, X., Tao, C., and Zhu, Y. (2013). Preparation and characterization of magnetic mesoporous bioactive glass/carbon composite scaffolds. *J. Chem.*, **2013**.
89. Andrade, A. L., Souza, D. M., Pereira, M. C., Fabris, J. D., and Domingues, R. Z. (2009). Synthesis and characterization of magnetic nanoparticles coated with silica through a sol–gel approach. *Cerâmica*, **55**, pp. 420–424.
90. Fagerlund, D. S., and Hupa, L. (2010). Crystallization of 45S5 during isothermal heat treatment. *Materialy Ceramiczne/Ceramic Materials*, **62**, pp. 349–354.

Chapter 2

Self-Assembly Approach for Biomaterials Development

Gunjan Verma and P. A. Hassan
Chemistry Division, Bhabha Atomic Research Centre, Mumbai-400085, Maharashtra, India
gunjanv@barc.gov.in

Self-assembly has evolved as a powerful tool to fabricate various nanostructured materials for biomedical applications. Self-assembly involves non-covalent interaction among constituent molecules and hence self assembled structures are susceptible to diverse structural transformations by changing external conditions such as pH, ionic strength, and solvent. Some of the fundamental principles of the self-assembly process in amphiphilic molecules are discussed in this chapter. The applications of self-assembly process in developing new materials with tailored properties have also been addressed. Direct application of the assembled structures as vehicles for drug delivery and as scaffolds for the synthesis of other inorganic materials has demonstrated its potential. Some of the important challenges that need to be addressed for extending this approach over a wide range of applications are also discussed.

Trends in Biomaterials
Edited by G. P. Kothiyal and A. Srinivasan
Copyright © 2016 Pan Stanford Publishing Pte. Ltd.
ISBN 978-981-4613-98-9 (Hardcover), 978-981-4613-99-6 (eBook)
www.panstanford.com

2.1 Phenomena: Principles of Self-Assembly

Self-assembly is a fundamental reversible process in which spontaneous organization of individual components into well-defined structures takes place under thermodynamic equilibrium conditions. This self-association of molecules forming hierarchical structures mostly involves a number of non-covalent interactions such as hydrogen bonds, electrostatic interaction, and van der Waals and hydrophobic forces. The self-assembly process can be classified as either static or dynamic. Static self-assembly involves the formation of ordered structures under equilibrium conditions without involving dissipation of energy. On the other hand, in the case of dynamic self-assemblies, the organization of disordered components requires dissipation of energy. These structures are described as "self-organized." Self-assembly is involved in producing structural organization on all scales ranging from molecules to galaxies. If the constitutive elements of self-assembled structures are molecules, the process is named "molecular self-assembly." Molecular self-assembly can be either intramolecular or intermolecular. In intramolecular self-assembly, the molecules assemble in such a way that from the random coil conformation, they change to well-defined stable structure. An example of intramolecular self-assembly is protein folding. In intermolecular self-assembly, the molecules associate themselves to form supramolecular assemblies, for example, formation of a micelle by amphiphilic molecules in solution [1–5].

Amphiphilic molecules, as the name suggests, contain both a hydrophilic and hydrophobic group, and to avoid the contact of their hydrophobic group with that of aqueous phase, they self-assemble at interfaces and in the bulk. Depending on the charge present at the head group of amphiphiles (surfactants), they have been classified into different categories: cationic, anionic, nonionic, and zwitterionic [6–8]. Figure 2.1 shows the schematic representation of a surfactant showing its hydrophilic and hydrophobic moieties and the examples for different categories of surfactants.

The self-assembly process is very common in naturally occurring biological molecules such as lipids as well as in synthetic amphiphiles, including surfactants, block copolymers, etc. The self-association of amphiphilic molecules occurs above a narrow range of concentration, which is known as critical micelle concentration

(CMC) and is a characteristic of each solvent–solute system. Below CMC, amphiphilic molecules remain as free individual molecules in solvent, and as the total concentration of amphiphilic molecules increases and exceeds CMC, the molecules begin to self-associate into nanoaggregates called micelles. These aggregates coexist with amphiphilic monomers in water. The formation of aggregates is accompanied with a sudden, well-defined change in some of the physicochemical properties such as interfacial tension, equivalent conductivity, light scattering, turbidity, osmotic pressure, self-diffusion coefficient, solubilization, and viscosity (Fig. 2.2) [9–11]. This well-defined change in different properties is ascribed to the formation of micellar aggregates and hence is used to determine the CMC of surfactant solutions.

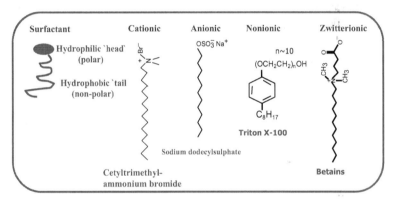

Figure 2.1 Schematic representation of a surfactant and examples of different types of surfactants classified based on their head groups.

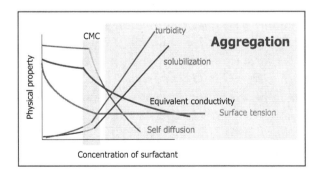

Figure 2.2 Schematic representation of different physical properties of surfactant solution showing critical micelle concentration.

In principle, any of the physical properties demonstrated in Fig. 2.2 can be used to find out the CMC of the surfactant; however, surface tension is the most commonly used technique for determining CMC. The CMC of each amphiphile is a characteristic feature at a given temperature and electrolyte concentration.

2.2 Historical Perspective

Self-assembly is omnipresent in nature in which the nanometer-scale building blocks spontaneously assemble into large-scale materials. For example, formation of silk material through monomeric silk fibroin protein is a very well-known classic example of the self-assembly process. Similarly, the phospholipid molecules having the length of a few nanometers can self-assemble into millimeter-size lipid tubules. Even ligaments and hair are the resultant of the self-assembly of collagen and keratin molecules. In addition to this, there are several naturally occurring examples in which the self-assembly process is involved in the fabrication of sophisticated structures and materials. Fabrication of mineralized structures is a widespread phenomenon among living organisms (e.g., shells, spines, spicules, bones, and teeth). In the formation of these structures, biomineralization takes place on the scaffold formed by self-assemblies. For example, in mammals, teeth form through biomineralization on a protein scaffold formed by the self-assembly of individual proteins.

Research on nanostructured materials created through self-assembly has gained increasing interest in pharmaceutical science. These materials have shown incredible potential in the field of therapeutics and diagnostics applications. A variety of colloidal drug carrier systems such as micellar solutions, hydrogels, polymeric microspheres, vesicles, liposomes, and cubosomes are being developed to be used as promising candidates for drug delivery systems. The prime objective of developing these formulations is to obtain systems with optimized drug loading and release properties, long shelf-life and low toxicity, and protection from chemical degradation. The designing of nature-inspired materials using molecular self-assemblies is also emerging as a new route to produce novel materials and to harmonize other materials such as ceramics, metals, and alloys. The use of self-assembled structures in the synthesis of templated inorganic minerals, such as silica, alumina, titania (titanium dioxide), and hydroxyapatite, has been a topic of great interest for many decades. The use of self-assembly as a template in synthesizing these materials provides several unique features such as controlled geometry, alignment, and porous structure. In particular, the synthesis of hydroxyapatite using molecular self-assembled structures has been of great interest owing to its similarity with the mineral constituents of human bones and teeth. In this chapter, we will mainly discuss the guiding rules for designing different types of self-assembled structures and their role in the development of biomaterials.

2.3 Design of Self-Assembled Structures

A variety of phases, including micelles, vesicles, micro-emulsions, and liquid crystalline dispersions, emerge by the self-association of amphiphilic molecules [12]. The geometry of these aggregates is a resultant of an interplay of various forces that drive the self-assembly. The "hydrophobic effect" of the hydrocarbon tails tends to bring the molecules closer together, and the "solvation" of the head groups tends to keep the hydrophilic part away from each other. The structure of aggregates broadly depends on the nature of the surfactant molecule, surfactant concentration, temperature, ionic strength of the solution, and nature of the additives. To predict the geometry of self-assembled structures formed in a surfactant–water system, a number of models have been proposed. The geometric packing models of Tanford [13] and Israelachivili and coworkers [14] suitably explain the shape of self-assemblies. As per their theory, the shape of the aggregates is decided by mainly three factors—the volume v of the hydrophobic part, the length l of the hydrophobic chain, and the effective head group area a_o of the surfactant molecule. The length and volume of the surfactant can be obtained by the Tanford formulae:

$$v = 27.4 + 26.9n \text{ Å}^3 \tag{2.1}$$

$$l = 1.5 + 1.265n \text{ Å} \tag{2.2}$$

Here, n represents the number of carbon atoms in the linear alkyl chain. The head group area per monomer, a_o, can be calculated from the Gibbs adsorption isotherm using the surface tension data. The geometries for the self-assembled structures formed in the solution depend on the allowed packing of surfactant molecules into aggregates, which can be easily described by a dimensionless parameter known as the "critical packing parameter $(v/a_o l)$." Hence, the value of the packing parameter is the deciding factor for the allowable geometries of the aggregates. Simple geometrical calculations have shown that the value of the packing parameter is different for the different shapes of the aggregates (Fig. 2.3). For example, for a spherical micelle of radius R and aggregation number N, the total volume of the micelle can be written as:

$$\frac{4}{3}\pi R^3 = Nv \tag{2.3}$$

and the total surface area of the micelle is:

$$4\pi R^2 = Na_o \tag{2.4}$$

As per the packing criteria, the radius of the spherical micelle cannot be more than the length l of the hydrocarbon tail, that is:

$$R = 3\frac{v}{a_o} \leq l \tag{2.5}$$

or in terms of the packing parameter $\dfrac{v}{a_o l} \leq \dfrac{1}{3}$.

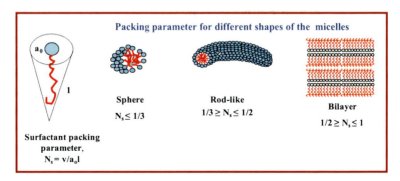

Figure 2.3 Schematic representation of different types of aggregates and their relation to the packing parameter of the surfactant.

Hence, if the packing parameter is less than 1/3, spherical micelles are the preferred form of aggregates. Similarly, cylindrical micelles are formed when the packing parameter is between 1/3 and 1/2, and highly curved bilayer vesicles are preferred when it is greater than 1/2. However, flat bilayers are formed as it approaches 1. For reverse micelles and without micro-emulsions, the value of the packing parameter is greater than 1.

Another parameter, known as the curvature energy, is also equally important in the case of vesicles, bilayers, etc., for dictating the structure of the supramolecular assemblies. According to the classical curvature model of Helfrich, the free energy per unit area of a bilayer associated with bilayer curvature can be written as:

$$\frac{F}{A} = \frac{1}{2}k_s(C_1 + C_2 - C_s)^2 \tag{2.6}$$

where c_1 and c_2 denote the two principal curvatures of the bilayer, and c_s denotes the spontaneous curvature. Here κ_s is the measure of the rigidity of the bilayer and is known as the bending modulus, which is of the order of $k_B T$. The spontaneous curvature arises from the packing considerations of the surfactant molecule itself. If a smaller packing area is favored due to the interaction between polar heads as compared to dictated by the tail–tail interactions, the surfactant monolayer will tend to curve in such a way that the polar regions are on the inside of the interface. On the other hand, if the head group packing prefers a larger area than that dictated by chains, the curvature will be such that the polar region is on the outside of the interface.

2.4 Characterization of Self-Assembled Aggregates

Scattering techniques are very promising tools for characterizing self-assembled structures. In scattering techniques, an incident radiation is scattered by the sample and we get the information about the size, shape, and orientation of the structures by analyzing the resulting scattering patterns. Depending on the radiation involved, these techniques have been classified into various categories; for example, if the incident radiation is light, the scattering technique is termed light scattering [static light scattering (SLS) as well as dynamic light scattering (DLS)]. Similarly, if the radiation involved is neutron or X-rays, the techniques are known as small-angle neutron scattering (SANS) and small-angle X-ray scattering (SAXS), respectively. Depending on the sample environment, the probable length scales, and the information to be obtained, the technique to be used is decided.

2.4.1 Light Scattering

Light scattering techniques are considered the most accessible techniques for the characterization of self-assembled structures. In a basic light scattering experiment, the average scattered intensity I_s (that is, the average rate of flow of energy per unit area of cross section) or the energy spectrum of intensity, $S(\omega)$, is measured [15]. The angular distribution of the scattered light intensity from a colloidal solution is a function of various factors such as size and shape of the colloidal particle, polarizability, and refractive index of the particles as well as the wavelength of the incident light. In a light scattering experiment, I_s bears information about the static properties of the scatterer, whereas $S(\omega)$ reflects the dynamics of the scatterer. Hence, in general, the light scattering experiments can be classified under two broad categories: SLS and DLS. In SLS, the average scattering intensity at various angles is measured assuming that the particles are at fixed positions in space or all orientations are considered from sufficient time average. However, in DLS, the energy spectrum of intensity or the time dependence of the intensity is measured taking into account the fact that the suspended particles are no longer stationary in reality; rather, they move about in a random walk fashion by a process known as Brownian motion. Due to the random movement of the scatterers, the phases of each of the scattered waves reaching the detector fluctuate randomly in time. As a result of this, the overall intensity fluctuates randomly in time due to the fact that the net intensity measured by the detector is a result of the superposition of all the waves scattered from the scattering volume [16]. These fluctuations in intensity depend on the size of the diffusing particles because small particles diffuse in the solution fast, resulting in a rapidly fluctuating intensity signal, whereas the larger particles, which diffuse slowly, result in slow fluctuations in intensity. The autocorrelation function, denoted by $C(\tau)$, represents the correlation or comparison between the values of the scattered intensity at a given time t and at a later time $t + \tau$. This comparison is made at different values of t to get a good statistical average. By representing the intensity at an arbitrary time as $I(o)$ and those at a later time τ as $I(\tau)$, the autocorrelation function can be written as:

$$C(\tau) = \langle I(o) \cdot I(\tau) \rangle \tag{2.7}$$

In the extreme limit in which τ becomes very large, there should not be any correlation between the pairs of sampled intensities, and hence the Eq. 2.7 reduces to:

$$C(\infty) = \langle I(o) \rangle^2 \tag{2.8}$$

An autocorrelator accepts the digital photo counts from the detector, which represents the light scattering intensity $I(t)$, and computes the second-order correlation function. The detector can be a photomultiplier tube coupled with a pulse amplifier and discriminator, or it can be an avalanche photodiode. The correlation function $C(\tau)$ is normalized with the long time correlation data, $C(\infty)$, that is equal to $<I>^2$. The normalized time correlation function of the scattered intensity $g^{(2)}(\tau)$ can be written as:

$$g^{(2)}(\tau) = \frac{<I(o)I(\tau)>}{<I>^2} \tag{2.9}$$

For photo counts obeying Gaussian statistics, $g^{(2)}(\tau)$ is related to the first-order correlation function of the electric field $g^{(1)}(\tau)$ by the Siegert relationship:

$$g^{(2)}(\tau) = a' + b\left|g^{(1)}(\tau)\right|^2 \tag{2.10}$$

where a' is the baseline and b is an adjustable parameter dependent on the scattering geometry and independent of τ.

For a suspension of monodisperse, rigid, spherical particles undergoing Brownian diffusion, the correlation function decays exponentially and is given as:

$$g^{(1)}(\tau) = (b)^{0.5} e^{-Dq^2\tau} \tag{2.11}$$

where D is the translational diffusion coefficient.

For small, dilute, non-interacting spheres, the hydrodynamic radius R_h can be obtained from the translational diffusion coefficient using the Stokes–Einstein relationship [17]:

$$D = \frac{kT}{6\pi\eta R_h} \tag{2.12}$$

where k is the Boltzmann constant, η is the solvent viscosity, and T is the absolute temperature. If the particle is non-spherical, R_h is often taken as the apparent hydrodynamic radius.

2.4.2 Neutron Scattering

Neutron scattering consists of a whole family of techniques, which give complementary information about the material. For example, coherent elastic scattering (diffraction) of neutrons gives the crystallographic or the magnetic structure of the material, whereas coherent inelastic scattering experiments determine phonon dispersion curves. On the other hand, incoherent inelastic and quasi-elastic measurements on hydrogenous samples give the nature of translational and/or rotational molecular motions in solids and liquids. The SANS technique is used for studying the structure of a material in the length scale of 10–1000 Å. This technique is known as SANS since the wavelength of neutrons used for the experiments is typically 4–10 Å, while the wave vector transfer (q) typically ranges between 0.001 and 1.0 Å^{-1}. Since such small q values occur at small scattering angles, the technique is called SANS. In SANS experiments, a beam of collimated, though not necessarily monochromatic, radiation is directed at a sample, and the flux of the radiation scattered into a solid angle element is recorded by a detector positioned at some distance and scattering. From the differential scattering cross section obtained by the SANS experiment, we get information about the shape, size, and interactions of the particles in the sample. The differential scattering cross section from a scattering volume consists of two terms—intraparticle scattering and interparticle scattering—and is represented as [18–20]:

$$\frac{d\Sigma}{d\Omega}(q) = n(\rho_p - \rho_m)^2 V^2 P(q) S(q) \tag{2.13}$$

where n is the number density of the particles; V is the volume of one scattering body; $(\rho_p - \rho_m)^2$ is the contrast factor; $P(q)$ is a function known as the form or shape factor, which depends on the shape, size, and polydispersity of the particle; $S(q)$ is the interparticle structure factor and is the function of interparticle interaction; and q is the magnitude of the scattering vector. The contrast factor in SANS experiments is equivalent to the difference in the refractive indices of the particle and the solvent in light scattering experiments. The scattering lengths of hydrogen (-0.3723×10^{-12} cm) and deuterium (0.6674×10^{-12} cm) are quite different; hence by deuterating either the particle or the solvent, a good contrast between the hydrogenous particle and the solvent can be made, which results in good scattering intensity.

2.4.3 Small-Angle X-Ray Scattering

Small-angle X-ray scattering also works on the same principle as that of SANS. In SAXS experiments, the structural information such as shape and size of macromolecules, size distribution, pore size, and characteristic distances of partially ordered materials, can be obtained from the intensity distribution of the scattered X-ray beam at very low scattering angles, typically in the range 0.1°–10° [21].

It is worth mentioning that all these techniques have their own advantages and limitations. For example, from light scattering, objects with the length scales of the order of the wavelength in the visible region of the electromagnetic spectrum can be probed. On the other hand, from SANS and SAXS, much smaller length scales can be probed. Light scattering and SAXS are easily accessible techniques; however, a special kind of instrumentation is required for SANS. Further, the SAXS intensity depends on the electron density, while SANS is sensitive to the neutron scattering cross section of the specimen, which makes it a promising tool for the samples having light elements. Also the possibility of isotope labeling and the strong scattering by magnetic moments make it a very advantageous technique.

In addition to scattering techniques, there are several other techniques such as microscopy and spectroscopy that help in elucidating the microstructure of self-assembled structures. In recent years, cryogenic transmission electron microscopy (Cryo-TEM) has become a powerful technique for the characterization of soft matters. A wide range of length scales ranging from a few nanometers to several micrometers can be imaged through Cryo-TEM. Also Cryo-TEM easily differentiates between the structures that are difficult to differentiate by scattering techniques, such as small disks and spherical micelles, or narrow ribbons and cylindrical micelles. Apart from structural characterization, having information regarding flow behavior of these aggregates helps in their effective utilization for different applications. Rheology is the technique that deals with the deformation and flow of matter in which the response of the materials is monitored when it is subjected to some external forces [22].

2.5 Structural Studies on Surfactant Assemblies

From the basic principles of self-assembly, it is evident that by manipulating the interactions between amphiphiles, the microstructure of the assemblies can be tuned, which offers promising prospects for the design of supramolecular materials for diverse applications. The properties (size, charge, etc.) of these carriers can be modified depending on the requirement for particular application. A variety of surfactant systems have been investigated in which structural transitions have been monitored with the addition of different additives such as electrolytes and hydrotropes. Various research groups have demonstrated such structural transitions in surfactant solutions using techniques such as SLS, DLS, SANS, and rheological studies. For example, DLS studies on a mixture of a cationic surfactant cetyltrimethylammonium bromide (CTAB) and anionic hydrotropic salt sodium p-toluene sulfonate (Na-PTS) demonstrate that the average hydrodynamic diameter of the micelles increases with increasing concentration of Na-PTS [23]. Figure 2.4 shows the hydrodynamic diameter of the CTAB micelles in the presence of different concentrations of Na-PTS as obtained by DLS studies.

Similar to that of cationic surfactants, in anionic surfactants also, structural transitions have been reported with the addition of cationic hydrotropes. For example, in an anionic surfactant, sodium dodecylsulfate (SDS), spherical to rod-like micellar transitions were investigated by the incorporation of a cationic hydrotrope, aniline hydrochloride (AHC), and its ortho- and meta-substituents—ortho-toluidine hydrochloride (OTHC) and meta-toluidine hydrochloride (MTHC)—by carrying out DLS and SANS studies [24]. The SANS spectra for 50 mM SDS solutions at room temperature in the presence of 20 mM AHC, OTHC, and MTHC are shown in Fig. 2.5.

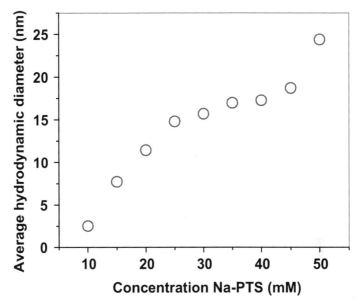

Figure 2.4 Average hydrodynamic diameter of CTAB micelles in the presence of different concentration of Na-PTS.

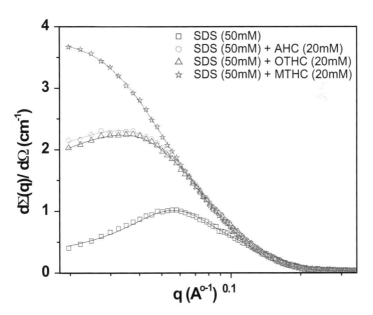

Figure 2.5 SANS spectra of pure SDS micelles in the presence of different hydrotropes.

The different features of the SANS spectrum give qualitative information about the system. For example, the presence of characteristic correlation peak in the SANS spectrum of pure SDS micelles indicates the presence of repulsive intermicellar interactions between the negatively charged SDS micelles. With the addition of salts, the correlation peak broadens, which is an indication of a decrease in the range of electrostatic interactions at constant volume fraction of the micelles. Also there is a shifting of the correlation peak to lower q values suggesting an increase in the value of intermicellar distance. In addition to this, the SANS spectra in the presence of all the salts merge in the high q region, which suggests that the micelles are growing uniaxially with the addition of different cationic hydrotropes. The solid lines in the figure represent the calculated scattering patterns obtained assuming prolate ellipsoidal micelles. The parameters obtained from the quantitative analysis show that the micelles grow uniaxially upon the addition of hydrotropes, and the growth of the micelles is much more pronounced with the addition of MTHC as compared to AHC and OTHC due to the more hydrophobic nature of MTHC as compared to that of AHC and OTHC.

Micelle-to-vesicle transitions have also been induced in surfactant assemblies by changing the nature and concentration of hydrotropic salts. For example, in a mixture of CTAB and an anionic hydrotrope, sodium hydroxylnapthalenecarboxylate (SHNC), having salt concentration more than that of surfactant, micelle–vesicle transitions have been investigated using several complementary techniques such as DLS, SANS, and rheology, upon simply diluting the solution [25]. The evolution of the SANS spectra for CTAB–SHNC mixtures at two different volume fractions is shown in Fig. 2.6.

Upon dilution, in the low q regime, the scattering pattern changes from a q^{-1} behavior to q^{-2}. In the SANS spectrum, the q^{-1} behavior is an indication of the presence of rod-like micelles, while the q^{-2} behavior suggests the presence of vesicles. Hence, a transition from rod-like micelles to vesicles could be monitored in CTAB–SHNC-mixed micelles by decreasing the volume fraction. The SANS data in Fig. 2.6 could be fitted well using rod-like micelles at high volume fraction (ϕ = 0.1), and the length of the micelles increases significantly on dilution. For ϕ = 0.000625, the data could be fitted well by considering a mixture of rods and unilamellar vesicles.

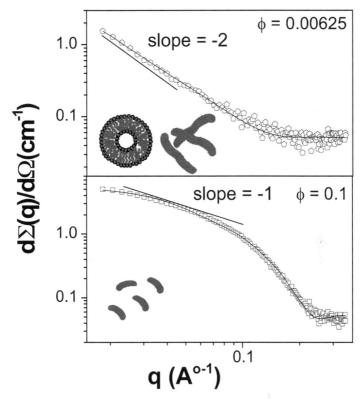

Figure 2.6 SANS spectra of CTAB–SHNC-mixed micelles at different volume fractions.

Recently, stimuli-responsive, particularly pH-sensitive, self-assembled structures have attracted great attention due to their versatile applications in pharmaceutical, biomedical, and related fields. As a model system, pH-responsive self-assemblies of CTAB—a cationic surfactant—and anthranilic acid (AA) were prepared. AA is hydrophobic in nature and possesses amine and carboxyl functional groups. The incorporation of AA into CTAB micelles makes them highly sensitive to external pH due to the presence of amine and carboxyl functional groups on AA [26]. Rheological, SANS, and DLS studies indicate that with increase in the pH of the CTAB–AA mixture, the micelles undergo elongation and form a three-dimensional network. Figure 2.7 shows the variation of viscosity of CTAB–AA-mixed micelles as a function of shear rate at different pH. At low pH, the viscosity is independent of shear rate, which is a characteristic of a Newtonian fluid and is comparable to that of pure water. With increase in the pH of the solution, the zero shear viscosity η_o (inset of Fig. 2.7) increases and reaches a maximum at around pH 5 due to the formation of sufficiently long micelles, which entangle with each other. In addition to this, a shear thinning is observed at high shear rate, which is a typical behavior of polymer-like micelles that undergo a reorientation of the anisotropic micelles when subjected to flow. On further increasing the pH of the mixture above 5, the viscosity decreases. Such decrease in viscosity beyond the maximum may be due to the transition from a linear micelle to branched network, or transition to a lamellar phase.

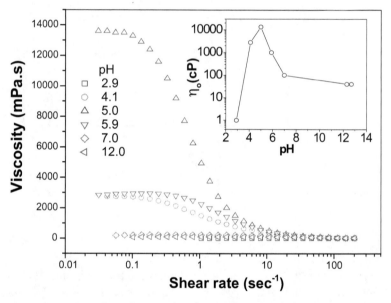

Figure 2.7 Variation of viscosity of CTAB–AA micelles as a function of shear rate at different pH. Inset shows η_o vs. pH plot.

This behavior of CTAB–AA micelles may be due to the fact that on increasing the pH of the solution, the carboxyl group dissociates and makes the AA molecules negatively charged. This negatively charged moiety decreases the effective repulsion between the positively charged head groups of CTAB. As a result, the surfactant packing parameter increases and it favors the growth of the micelles.

Similar to that of cationic surfactant, AA molecules could be incorporated in the micelles of nonionic surfactant Triton X-100, and pH-induced changes in the surface charge of nonionic micelles have also been demonstrated using different techniques [27]. The structure and dynamical properties of block copolymers have also been tuned by the incorporation of different additives. For example, the addition of a hydrophobic nonionic surfactant Surfynol®104 in the solution of a series of Pluronics copolymer with varying hydrophile–lipophile balance (HLB) values not only enhances micellization but also promotes shape transition such as cylindrical micelles as well as bilayers in micelles [28, 29].

2.6 Applications

2.6.1 Self-Assemblies for Drug Delivery

Self-assembled materials formed of biocompatible surfactants and block copolymers have attracted rising interest in pharmaceutical science due to their incredible potential in the field of drug delivery applications. A variety of self-assembled structures such as micellar solutions, micro-emulsions, vesicle and liquid crystalline structures, microcapsules, liposomes, cubosomes, and colloidosomes have been developed for drug delivery applications with the aim to overcome the drawbacks of conventional drug carriers. These new generation drug delivery systems have been designed to minimize drug degradation, prevent harmful side effects, and increase drug bioavailability. Figure 2.8 shows some of the self-assembled structures formulated for drug delivery applications.

Polyoxyethylene-based nonionic surfactants such as Polysorbate 80 (Tween-80), Cremophor EL, and Brij-35 have been extensively used in various pharmaceutical preparations due to their high biocompatibility. Valium® MM and Konakion® MM are two mixed-micelle-based formulations currently available in the pharmaceutical market. The mixed micelles formed of Tween-80 and sodium deoxycholate (NaDC) [30] have been explored for the delivery of anticancer drugs such as doxorubicin hydrochloride (DOX). Cytotoxicity studies on different model cancer cell lines using this mixed micellar formulation showed two- to tenfold lower

LC$_{50}$ value as compared to that of plain DOX. Tween-80-based micellar formulations [31] have shown enhanced permeation of hydrocortisone through the cornea, as compared to non-micellar drug for the treatment of eye inflammation. In recent years, block-copolymer-based micelles are being regarded as an ideal tool for drug delivery due to their unique features such as high stability, low CMCs, as well as high versatility in composition. Micelles composed of Pluronic F127 have been employed for the transport of sumatriptan (an antimigraine drug) to brain through intranasal route [32]. *In vivo* studies were carried out using ^{99m}Tc as a radiolabeling agent for sumatriptan. A significantly higher brain uptake of sumatriptan-loaded micellar formulation was detected by biodistribution and autoradiography studies in rats as compared to that of sumatriptan solution. pH-responsive drug carrier systems for the targeted drug delivery of anticancer drug doxorubicin to human breast carcinoma MCF-7 cells have been developed using polyHis-*b*-poly(ethylene glycol)-folate and poly(L-lactic acid)-*b*-PEG-folate [33]. Oral route is one of the most favorable methods for drug administration. However, due to several obstacles in the gastrointestinal tract, it cannot be used for all drugs. The group of Kabanov has made significant contributions to oral drug delivery using Pluronic® triblock copolymer micellar formulations [34–36].

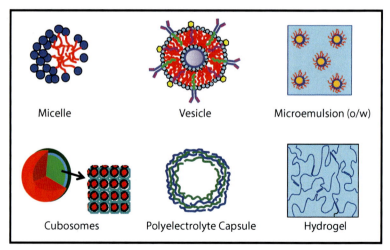

Figure 2.8 Schematic representation of some of the self-assembled drug delivery systems.

Vesicles are another class of self-assembled structures used for drug delivery applications. They are hollow spherical, layered structures formed by the self-assembly of surfactants, lipids, or block copolymers in aqueous solution. Depending on the number of bilayers, they can be classified as unilamellar and multilamellar vesicles. If vesicles are formed of many concentric spherical bilayers of phospholipids, they are known as liposomes, and depending on the functionalities present, they can be classified as pH-sensitive liposomes, magnetic liposomes, niosomes, and polyhedral niosomes. There are several reports in which liposomes have been used as drug delivery vehicles. A pH and temperature dual-sensitive liposome gel using methoxypolyethylene glycol 2000-hydrazone-cholesteryl hemisuccinate polymer was formulated containing arctigenin for the treatment of vaginal candidiasis [37]. Multilamellar vesicles of a biodegradable surfactant PEG-8 Distearate (PEG8DS) were employed for the encapsulation of sumatriptan succinate, an antimigraine drug [38]. Liposomes have been approved as a carrier for the delivery of various drugs, such as amphotericin B, cytaribine, daunorubicin, doxorubicin, and morphine. DAUNOXOME®, DOXIL®, and AmBisome® are the commercial liposome-based drugs available in the market.

Cubosomes are nanostructured particles formed by high-energy dispersion of cubic liquid crystals followed by a stabilization using polymeric surfactants [39]. Cubosomes have several advantages compared to that of their parent liquid crystalline phase such as larger specific surface area, much lower viscosity, and stability at any dilution level. The most common surfactant used to make cubosomes is the monoglyceride glycerol monoolein. Liquid crystalline particles composed of reverse hexagonal phases are termed hexosomes. Cubosomes have several unique features such as ability to incorporate both hydrophilic and hydrophobic drugs, biologically compatible microstructure, and capability of controlled release of solubilized active ingredients such as drugs and proteins, which makes them interesting candidate for drug delivery applications. For example, cubosome particles of monoolein and water in the presence of stabilizer poloxamer 407 loaded with dexamethasone (DEX) showed a significant increase in the apparent permeability coefficient compared to that of DEX-sodium phosphate eye drops through *in vitro* studies [40]. Also the retention of cubosomes was significantly longer than that of solution and carbopol gel without affecting the corneal structure and tissue integrity, indicating the good biocompatibility of DEX cubosomes.

Colloidosomes are the microcapsules fabricated by the self-assembly of colloidal particles on to the interface of a core. The core of the colloidosomes may be composed of a polymer gel consisted of a cross-linked network of long polymer molecules, while the outer shell can be composed of a semiconductor, magnetic material, polymer, or an inorganic material. Colloidosomes are classified based on the composition of core and shell. For example, depending on the gel forming the core, they are termed an aqueous or oily gel core colloidosomes. Similarly, based on the self-assembled colloidal particles in the shell, they are named nanoparticle colloidosomes, magnetic colloidosomes, layer-by-layer (LbL) colloidosomes, hairy colloidosomes (shell consisting of rod-shaped particles), and pH-responsive colloidosome [41–45]. Colloidosomes with porous shells are of great importance for drug delivery applications. Spray-dried hybrid microparticles of Capmul MCM (mono-diglycerides of C_8/C_{12} fatty acids) and silica (Aerosil® 380) were found to show significant enhancement in the oral bioavailability of a non-steroidal anti-inflammatory drug, celecoxib, compared to an aqueous suspension after oral administration to beagle dogs [46].

Tubules are another class of self-assembled structures formed of rolled-up bilayer sheets of amphiphilic molecules such as diacetylenic lipids, amide amphiphiles, and diblock copolymers. They have hollow cylindrical structures with opened ends and an internal aqueous space and can be of variable diameters, lengths, and wall thicknesses depending on the nature of amphiphiles and the preparation conditions. Self-assembled peptide nanotubes and microtubes have attracted a lot of interest for drug delivery applications. Cyclic peptide nanotubes composed of (Trp-D-Leu)4-Gln-D-Leu have been demonstrated to be used for the delivery of antitumor drug 5-fluorouracil [47]. The microtubes composed of bolaamphiphile, bis(N-α-amido-threonine)-1,5-pentane dicarboxylate have been shown as pH- and concentration-dependent drug release vehicles for ellagic acid (EA) [48].

Polyelectrolyte capsules are hollow LbL assembly of synthetic polymers formed by the adsorption of oppositely charged synthetic polymer layers around a charged spherical core. Synthetic anionic poly(sodium) styrene sulfonate and cationic poly(allylamine) hydrochloride are the most widely used polyelectrolyte pairs for the formation of polyelectrolyte capsules [49]. The core template of the capsules should be stable under the LbL capsule formation process, soluble in mild conditions, and also should be completely removable from the inside of the capsules after completion of LbL adsorption process. It is also possible to tune the properties of the capsule wall, such as thickness, permeability, porosity, and elasticity, as they depend on the chemical structure and the molecular weight of the polyelectrolyte pairs used for the fabrication of capsules. In addition to this, different types of functionalities such as fluorescent nanoparticles, metal nanoparticles, magnetic nanoparticles, as well as to enhance the biocompatibility of the capsules, biological molecules may also be incorporated into the walls [50–53]. Recently, a polyelectrolyte multilayer capsule filled with bovine serum albumin gel (BSA-gel-capsule) showed excellent pH-controlled loading and release properties and intravenous administration of fluorescein isothiocyanate-labeled BSA-gel-capsules into mice showed a remarkable targeting action to the lung [54].

Micro-emulsions are thermodynamically stable colloidal dispersions of water and oil stabilized by an interfacial film of surfactant or a combination of surfactant and a co-surfactant, which is known as stabilizer. Micro-emulsions can be categorized as oil-in-water (o/w), water-in-oil (w/o), and bicontinuous micro-emulsions. Micro-emulsions have distinctive features such as high solubilization power as well as the ability to solubilize both hydrophilic and lipophilic drugs, long shelf-life, etc., which makes them promising for drug delivery applications. Several micro-emulsion-based formulations are currently in the market. For example, a micro-emulsion-based formulation composed of fine dispersion of cyclosporine A is marketed under the trade mark Neoral® by Novartis. This formulation gives rapid absorption and increased bioavailability, faster onset of action, and a reduced inter–intra patient variability as compared to earlier crude oil-in-water emulsion of cyclosporin, under the trade mark Sandimmune® [55]. Another example of micro-emulsion-based formulation is Solvium, which is a topical formulation of a poorly soluble drug, ibuprofen (IBU), into a transparent gel marketed by Chefaro (Akzo).

Hydrogels are three-dimensional networks of cross-linked hydrophilic polymer chains prepared by the entanglement of long fibrous polymeric structures in water. The entanglement of fibrous structures can be made by various methods such as chemical cross-linking by covalent bonds, hydrogen bonding, van der Waals interactions, physical entanglements, photopolymerization or irradiated cross-linking of synthetic and natural polymers. Hydrogels have the ability to swell in water by entrapping a large amount of water. Since hydrogels are formed by weak interactions, they show strong responses under the influence of external stimuli such as temperature, electric fields, solvent composition, light, pressure, sound and magnetic as well as pH, and specific molecular recognition. Stimuli-responsive hydrogels have demonstrated great potential for drug delivery applications. A temperature-sensitive hydrogel composed of quaternized chitosan and poly(ethylene glycol) containing a small amount of α-β-glycerophosphate (α-β-GP) has been tested for nasal insulin delivery [56]. The results showed that after the administration of hydrogel formulation, no cytotoxicity could be monitored and the blood glucose level decreased to 40–50% of initial blood glucose concentration for at least 4–5 h. A pH/temperature-sensitive pentablock copolymer of poly(beta-amino ester)-poly(epsilon-caprolactone)-poly (ethylene glycol)-poly(epsilon-caprolactone)-poly(beta-amino ester) hydrogel has been evaluated as a sustained injectable insulin delivery system on male Sprague–Dawley rats [57]. Upon subcutaneously injecting the complex mixtures of insulin-gel, the streptozotocin (STZ) diabetic rat could be treated for more than 1 week. In addition to pH- and temperature-sensitive hydrogels, other stimuli-responsive gels have also been employed for drug delivery applications. A nanocomposite of a temperature-sensitive hydrogel made of poly (N-isopropylacrylamide) and superparamagnetic Fe_3O_4 particles has been reported to show remote-controlled release of methylene blue and vitamin B_{12} by pulse application of a high frequency alternating magnetic field [58]. Upon the application of magnetic field, Fe_3O_4 particles warm up and accordingly the surrounding gel also gets heated up. As a result of this, depending on the application of magnetic field, the gel expands and contracts and regulates the release of the drug.

2.6.2 Self-Assembly in Biomineralization

Fabrication of sophisticated mineralized structures involving self-assembly process is a universal phenomenon among living organisms. There are several naturally occurring examples in which self-assembly is involved in the development of sophisticated structures and materials, for example shells, spines, spicules, bones, and teeth. In these biological structures, biomineralization takes place on a protein scaffold formed by the self-assembly of individual proteins. Although the components of these materials are very soft and brittle, the organic and inorganic constituents interact at the molecular and micro-levels synergistically and produce resultant structures with surprising mechanical properties. In the recent years, a great deal of attention has been paid to fabricate biomaterials using molecular self-assemblies to mimic nature-inspired materials. Also substantial advances have been made to produce potential biological materials for a wide range of applications using peptides, phospholipids, and DNA as building blocks.

As discussed in the earlier section of this chapter, self-assembly is an amazing phenomenon that involves the structuring of amphiphiles and solvent molecules. Having an understanding of self-assembly process regarding the CMC of the surfactant, Krafft temperature, variables that influence the structure of aggregates such as concentration, temperature, pH, etc., a better control can be set up over the final morphology and properties of the materials. Surfactant assemblies are dynamic structures that are very sensitive to the changes made in the surrounding solutions. For example, the addition of more surfactant can increase micelle size or change it to even liquid crystalline phase. The addition of inorganic electrolytes to ionic micelles significantly changes the size and shape of the aggregates by screening the charge at the surface of the micelles, which in turn changes the packing parameter. Organic compounds such as aromatic or aliphatic hydrocarbons dissolve in the hydrophobic cores of some micelles up to a certain concentration and cause micelle cores to swell. Hence, by changing the parameters influencing the microstructure of self-assembled structures, it is possible to bring out flexibility in the synthesis conditions and accordingly tune the morphology and porosity of the materials [23–29].

For many decades, the involvement of molecular self-assemblies as a template in creating the materials for diverse applications has been a topic of immense interest. In particular, a great deal of attention has been paid to the synthesis of hydroxyapatite using a variety of molecular self-assemblies due to its similarity with the mineral constituents of bones and teeth [59, 60]. Hydroxyapatite is a bioactive material; hence it supports bone ingrowth and osseointegration when used in orthopedic, dental, and maxillofacial applications. Also porous hydroxyapatite can incorporate biologically active molecules such as drugs and osteogenic agents that can treat bone infection and diseases and promote bone tissue regeneration [61–65]. Hydroxyapatite is an inorganic biocompatible material composed of calcium, phosphate, and hydroxide. It has the chemical formula $Ca_5(PO_4)_3(OH)$, which is commonly written as $Ca_{10}(PO_4)_6(OH)_2$, signifying that the crystal unit cell involves two entities. It crystallizes in the hexagonal crystal system. Bone is a hybrid structure of inorganic–organic material. The inorganic composition of bone is formed of needle-shaped carbonated and calcium-deficient hydroxyapatite nanocrystals. The organic part of the bone is mainly composed of collagen matrix, which dictates the properties of hydroxyapatite by controlling its nanoscale structure.

The preparation of synthetic hydroxyapatite has been a topic of interest for decades due to its excellent biocompatibility, bioactivity [66], and high osteogenic potential. Several groups have put their efforts in synthesizing hydroxyapatite with controllable properties and porous structure using a variety of techniques such as solid-state route, sol–gel method, hydrothermal reaction, precipitation method, mechano-chemical, emulsion technique, as well as by using surfactant assemblies as templates [67–70]. Analogous to that of collagen matrix, the involvement of synthetic polymers and surfactant assemblies in the synthesis of hydroxyapatite could mimic biomineralization and result in the materials with distinctive morphology and porosity, imparting them new functional properties. Stupp and coworkers recommended that biomolecules such as proteins, glycoproteins, and polysaccharides play an important role in controlling the nucleation and growth of mineral phases and synthesized apatite-based materials in aqueous solutions of homopolymer poly(amino acids) and synthetic polyelectrolytes [71]. They also developed a series of peptide amphiphiles that can self-assemble into well-ordered nanofibers [72, 73] and employed these nanofibers as nucleating center for the mineralization of hydroxyapatite. Self-assemblies involving ionic and nonionic surfactant as well as block copolymers have been employed as templates for the synthesis of nanostructured hydroxyapatite and have been shown to guide the morphology of the resulting hydroxyapatite particles. In the presence of spherical micelles of cationic surfactant cetrimide, spherical particles of hydroxyapatite were formed [74]. Similarly, Zhang and coworkers [75] have shown that in the presence of a nonionic surfactant Triton X-100, well-aligned hydroxyapatite nanorods are formed, templated by hexagonal phase of Triton X-100. In the presence of different concentration of dodecyl phosphate, nanocrystalline hydroxyapatite powder was synthesized and it was concluded that the surfactant concentration used during synthesis dictates the final properties of the hydroxyapatite nanoparticles [76]. A recent study has reported the synthesis of hydroxyapatite nanoparticles using rod-like micelles of a cationic surfactant, cetyltrimethylammonim bromide (CTAB), and an anionic hydrotrope, sodium salicylate (SS), as a templating agent and the comparison of its morphology and porosity behavior with respect to the hydroxyapatite prepared in the absence of any template under similar experimental conditions. The microstructural study showed that large hydroxyapatite nanorods of diameter about 50–100 nm are formed in surfactant-mediated synthesis, whereas irregular-shaped nanoaggregates of hydroxyapatite are obtained in the absence of surfactant. The presence of bimodal distribution of mesopores was evident from porosity measurements in hydroxyapatite prepared in the presence of surfactant template, while hydroxyapatite prepared in the absence of surfactant template shows monomodal distribution of mesopores. Thus, rod-like micelles composed of surfactants or other organic materials are found to be promising soft templates for the crystallization [77].

Synthetic hydroxyapatite has been widely applied in biomedical applications including bone tissue regeneration, cell proliferation, and drug delivery for many years due to its excellent biocompatibility, bioactivity and similarity in composition to that of bone material. In addition to chemical composition and structure for bone-forming ability of hydroxyapatite, the textural properties such as pore size, pore volume, pore structure etc. are also crucial factors for an excellent osteointegration. The porous nature of hydroxyapatite allows the ingrowth of bone tissue, and a full integration with the living bones can be attained, which provides a mechanical interconnect leading to a compact fixation of the material.

Porous hydroxyapatite has been extensively applied for repairing, regeneration, and reconstruction of lost, damaged, or degenerative bones. In the case of injuries where bone damage is too much and healing through autoregeneration of bones is difficult, bone grafts are required. In such cases, it is very important to match the osteoconductive properties of the material used for bone grafting to that of living bones. Hydroxyapatite has a capability to support bone ingrowth through osteoconduction mechanism without imparting any toxicity inflammation or foreign body response. Hydroxyapatite is also used as a filler to replace amputated bone and reconstruction of damaged bones and tooth zones. Due to its biocompatible, bioactive, and osteoconductive properties, hydroxyapatite may be employed for filling bone defects or voids. This bone filler works as a scaffold and promotes the fast filling of the voids by ingrowth of naturally occurring bone and offers an alternative to bone grafts. In this process, the scaffold becomes a part of the bone structure and reduces the healing time. In addition to this, there are several reports in which hydroxyapatite coating is applied to prosthetic implants [78, 79]. The coating with hydroxyapatite provides biocompatibility and bioactivity to metallic implants and hence the body does not consider it a foreign material.

In recent years, much attention has been paid to porous hydroxyapatite as a potential candidate for bone drug delivery system due to their physicochemical and biological properties. Several groups have designed porous hydroxyapatite for drug delivery applications using self-assembled structures as a template. For example, Li *et al.* synthesized alendronate-functionalized mesoporous nanosized hydroxyapatite using the cationic surfactant CTAB as a template and evaluated them for drug delivery applications using IBU as a model drug [80]. The alendronate-functionalized hydroxyapatite resulted in small particle size and high surface area and showed high IBU storage capacity and relatively slower release rate compared with pure hydroxyapatite due to the ionic interaction between $-NH_3^+$ on the matrix and $-COO^-$ on IBU. Mesoporous hydroxyapatite-silica (HA-silica) composites prepared through the sol–gel process using the triblock copolymer Pluronics P123 as template showed promising biocompatibility, and osteosarcoma cells were found to grow on the composites. The drug loading and sustained release capacity of the composites were dependent on the porosity of the samples [81]. The drug storage/release test on a europium-doped luminescent and mesoporous hydroxyapatite prepared using cationic surfactant as a template indicates that the luminescent hydroxyapatite shows much similar drug loading amount and cumulative release rate to those of pure hydroxyapatite using IBU as a model drug. In addition to this, the samples show red luminescence of Eu^{3+} ($^5D_0$–$^7F_{1,2}$) under UV irradiation even after loading IBU, and the emission intensities of Eu^{3+} in the drug carrier system vary with the released amount of IBU. This study indicates that the drug release can be easily tracked and monitored by the change of the luminescence intensity [82]. *In vitro* drug release experiments in PBS (pH 7.4) on a hydroxyapatite-coated poly(lactic-co-glycolic acid) microspheres prepared through the micro-emulsion method having an encapsulated antibiotic, amoxicillin showed a sustained release profile for at least 31 days with little initial burst release [83]. Palazzo *et al.* [84] prepared porous hydroxyapatite having a bimodal porosity and tested them for controlled drug delivery using ibuprofen-lysine and hydrocortisone Na-succinate as non-steroidal and steroidal anti-inflammatory drug-models, respectively. Composites of hydroxyapatite with collagen were developed as drug delivery system for an anticancer drug, paclitaxel (Tax). The Tax-loaded hydoxyapatite was found to be more toxic towards highly metastatic MDA-MB-231 cells compared to that of poorly metastatic MCF-7 cells. On the other hand, the free Tax-containing collagen gel did not show any *in vitro* drug actions against both of the cancer cells. Hence, this study suggests that hydroxyapatite could be used as an effective carrier system for the delivery of hydrophobic anticancer drugs, and that the drug-loaded hydroxyapatite-embedded collagen gel would be useful for cancer chemotherapy [85]. pH-responsive hydroxyapatite nanoparticles with a hollow core and a mesoporous shell (hmHANPs) synthesized by an opposite ion core/shell strategy exhibited superior drug loading capacity and enhanced drug release efficacy for an anticancer drug, doxorubicin, compared to that of hydroxyapatite nanoparticles without a hollow structure [86].

2.7 Future Directions

From the preceding discussions, it is evident that self-assembled structures offer an attractive route for encapsulation and controlled delivery of drugs due to their remarkable advantages such as improved drug solubility, enhanced bioavailability, and passive targeting. Although a variety of self-assembled materials are being formulated for drug delivery applications, it is worth mentioning that each of these carrier types has its own benefits and constraints. For example, most of these self-assembled structures are sensitive to dilution, which limits their usage during practical applications. However, by involving amphiphiles having ultra-low critical aggregate concentrations for the fabrication of assemblies, these aggregates can be stabilized to some extent. To limit the dilution-induced disassembly of self-assembled structures, they can also be incorporated in the hydrogel matrix having mesh size smaller than the dimensions of the aggregates. In addition to this, partial cross-linking of the surfactant monomers either at the core or at the shell can lead to the stabilization of the aggregate structure. The toxicity of these materials is another issue for their usage, which is linked to the chemical composition of the monomer units and the size and shape of the carriers. The cytotoxicity issues can be tackled by using lipids, bile salts, and other biocompatible amphiphiles. In another promising strategy, the drug molecules could be chemically bound with an amphiphile itself that can release under the influence of an external stimulus. These amphiphiles are known as prodrug amphiphiles. As it is very much convenient to modulate the microstructure of these materials by modulating the intermolecular interactions, it provides an opportunity to incorporate special functionalities in these materials for particular applications, for example, by incorporating pH- and temperature-responsive functional groups in amphiphilic molecules, the self-assemblies can be made stimuli responsive so that they can release drugs at specific conditions. Site-specific targeting can also be achieved by including targeting moieties such as folic acid and antibodies in the amphiphilic molecules.

In recent years, attempts have been made to understand and mimic the hierarchical structure of mineralized tissues using the self-assembly approach. The interest in biomineralization arises from their potential use in repairing of bone fractures, defects in the coating on a load-bearing metallic implant, or as a bone drug delivery vehicle for treating bone infection and diseases. A variety of templates have been employed during biomineralization for understanding and mimicking the structure of mineralized tissues. There are several reports, which show that the morphology of the template nanostructures may effectively control the nucleation and growth of the mineral. However, the relationship between the microstructure of the template and the orientation of mineral is not understood to a great extent, and the role of self-assembly mechanisms in controlling the dimensions and shapes of the mineralized inorganic materials remains a challenge.

References

1. Degiorgio, V.; and Corti, M. (1985) *Physics of Amphiphiles: Micelles, Vesicles and Microemulsions* (North-Holland, Amsterdam).
2. Zana, R. (1987) *Surfactant Solutions: New Methods of Investigation* (Marcel Dekker, New York).
3. Evans, D. F.; and Wennerström, H. (1999) *The Colloidal Domain* (Wiley-VCH, New York).
4. Nnanna, I. A.; and Xia, J. (eds.) (2001) *Protein-Based Surfactants*, Surfactant Science Series, Vol. 101 (Marcel Dekker, New York).
5. Hunter, R. J. (2005) *Foundations of Colloid Science* (Oxford University Press, New York).
6. Rosen, M. J. (2004) *Surfactants and Interfacial Phenomena* (John Wiley, New Jersey).
7. Myers, D. (2006) *Surfactant Science and Technology* (John Wiley, New Jersey).
8. Evans, D. F.; and Wennerstrom, H. (1994) *The Colloidal Domain: Where Physics, Chemistry, Biology, and Technology Meet* (VCH, New York).
9. Domínguez, A.; Fernández, A.; González, N.; Iglesias, E.; and Montenegro, L. (1997) Determination of critical micelle concentration of some surfactants by three techniques, *J. Chem. Educ.*, **74**, pp. 1227–1231.
10. Kunieda, H.; and Shinoda, K. (1976) Krafft points, critical micelle concentrations, surface tension, and solubilizing power of aqueous solutions of fluorinated surfactants, *J. Phys. Chem.*, **80**, pp. 2468–2470.

11. Holmberg, K.; Jonsson, B.; Kronberg, B.; and Lindman, B. (2003) *Surfactants and Polymers in Aqueous Solution*, 2nd ed. (John Wiley and Sons, New York).
12. Clint, J. H. (1991) *Surfactant Aggregation* (Chapman and Hall, New York).
13. Tanford, C. (1980) *The Hydrophobic Effect: Formation of Micelles and Biological Membranes*, 2nd ed. (John Wiley and Sons, New York).
14. Israelachvili, J. N. (1991) *Intramolecular and Surfaces Forces*, 2nd ed. (Academic Press, New York).
15. Sood, A. K. (1991) *In Solid State Physics, Advances in Research and Applications* (Academic Press, New York).
16. Pecora, R. (1985) *Dynamic Light Scattering: Application of Photon Correlation Spectroscopy* (Plenum Press, New York).
17. Einstein, A. (1956) *Investigations in the Theory of the Brownian Movement* (Dover, New York).
18. Glatter, O.; and Kratky, O. (1982) *Small Angle X-ray Scattering* (Academic Press, London).
19. Kaler, E. W. (1988) Small angle scattering from colloidal dispersions, *J. Appl. Cryst.*, **21**, pp. 729–736.
20. Pedersen, J. S. (1997) Analysis of small angle scattering data from colloids and polymer solutions: Modeling and least-squares fitting, *Adv. Colloid Interface Sci.*, **70**, pp. 171–210.
21. Guinier, A.; and Fournet, G. (1955) *Small Angle Scattering of X-rays* (Wiley Interscience, New York).
22. Barnes, H. A.; Hutton, J. F.; and Walters, K. (1989) *An Introduction to Rheology* (Elsevier, Amsterdam).
23. Garg, G.; Hassan, P. A.; and Kulshreshtha, S. K. (2006) Dynamic light scattering studies of rod-like micelles in dilute and semi-dilute regime, *Colloids Surf. A: Physicochem. Eng. Asp.*, **275**, pp. 161–167.
24. Garg, G.; Hassan, P. A.; Aswal, V. K.; and Kulshreshtha, S. K. (2005) Tuning the structure of SDS micelles by substituted anilinium ions, *J. Phys. Chem. B*, **109**, pp. 1340–1346.
25. Verma, G.; Aswal, V. K.; Fritz-Popovski, G.; Shah, C. P.; Kumar, M.; and Hassan, P. A. (2011) Dilution induced thickening in hydrotrope-rich rod-like micelles, *J. Colloid. Interface Sci.*, **359**, pp. 163–170.
26. Verma, G.; Aswal, V. K.; and Hassan, P. A. (2009) pH-Responsive self-assembly in an aqueous mixture of surfactant and hydrophobic amino acid mimic, *Soft Matter*, **5**, pp. 2919–2927.

27. Verma, G.; Aswal, V. K.; Kulshrestha, S. K.; Hassan, P. A.; and Kaler, E. W. (2008) Adsorbed anthranilic acid molecules cause charge reversal of nonionic micelles, *Langmuir*, **24**, pp. 683–687.
28. Thummar, A. D.; Sastry, N. V.; Verma, G.; and Hassan, P. A. (2011) Aqueous block copolymer–surfactant mixtures-surface tension, DLS and viscosity measurements and their utility in solubilization of hydrophobic drug and its controlled release, *Colloids Surf. A: Physicochem. Eng. Asp.*, **386**, pp. 54–64.
29. Kadam, Y.; Bharatiya, B.; Hassan, P. A.; Verma, G.; Aswal, V. K.; and Bahadur, P. (2010) Effect of an amphiphilic diol (Surfynol®) on the micellar characteristics of PEO–PPO–PEO block copolymers in aqueous solutions, *Colloids Surf. A: Physicochem. Eng. Asp.*, **363**, pp. 110–118.
30. Bhattacharjee, J.; Verma, G.; Aswal, V. K.; Date, A. A.; Nagarsenker, M. S.; and Hassan, P. A. (2010) Tween 80-sodium deoxycholate mixed micelles: Structural characterization and application in doxorubicin delivery, *J. Phys. Chem. B*, **114**, pp. 16414–16421.
31. Zimmer, A. K.; Maincent, P.; Thouvenot, P.; and Kreuter, J. (1994) Hydrocortisone delivery to healthy and inflamed eyes using a micellar polysorbate 80 solution or albumin nanoparticles, *Int. J. Pharmaceutics*, **110**, pp. 211–222.
32. Jain, R.; Nabar, S.; Dandekar, P.; Hassan, P. A.; Aswal, V. K.; Talmon, Y.; Shet, T.; Borde, L.; Ray, K.; and Patravale, V. (2010) Formulation and evaluation of novel micellar nanocarrier for nasal delivery of sumatriptan, *Nanomedicine*, **5**, pp. 575–587.
33. Lee, E. S.; Na, K.; and Bae, Y. H. (2005) Super pH-sensitive multifunctional polymeric micelle, *Nano Lett.*, **5**, pp. 325–329.
34. Alakhov, V.; Klinski, E.; Lemieux, P.; Pietrzynski G.; and Kabanov, A. (2001) Block copolymeric biotransport carriers as versatile vehicles for drug delivery, *Expert. Opin. Biol. Ther.*, **1**, pp. 583–602.
35. Kabanov, A.; Batrakova, E.; and Alakhov, V. (2002) Pluronic® block copolymers as novel polymer therapeutics for drug and gene delivery, *J. Control. Release*, **82**, pp. 189–212.
36. Kabanov, A. V.; Batrakova, E. V.; and Alakhov, V. Y. (2002) Pluronic® block copolymers for overcoming drug resistance in cancer, *Adv. Drug. Del. Rev.*, **54**, pp. 759–779.
37. Chen, D.; Sun, K.; Mu, H.; Tang, M.; Liang, R.; Wang, A.; Zhou, S.; Sun, H.; Zhao, F.; Yao, J.; and Liu, W. (2012) pH and temperature dual-sensitive liposome gel based on novel cleavable mPEG-Hz-CHEMS polymeric vaginal delivery system, *Int. J. Nanomed.*, **7**, pp. 2621–2630.

38. Redkar, M.; Hassan, P. A.; Aswal, V.; and Devarajan, P. (2007) Onion phases of PEG-8 distearate, *J. Pharm. Sci.*, **96**, pp. 2436–2445.
39. Fong, C.; Weerawardena, A.; Sagnella, S. M.; Mulet, X.; Waddington, L.; Krodkiewska, I.; and Drummond, C. J. (2010). Monodisperse nonionic phytanyl ethylene oxide surfactants: High throughput lyotropic liquid crystalline phase determination and the formation of liposomes, hexosomes and cubosomes, *Soft Matter*, **6**, pp. 4727–4741.
40. Gan, L.; Han, S.; Shen, J.; Zhu, J.; Zhu, C.; Zhang, X.; and Gan, Y. (2010) Self-assembled liquid crystalline nanoparticles as a novel ophthalmic delivery system for dexamethasone: Improving preocular retention and ocular bioavailability, *Int. J. Pharm.*, **396**, pp. 179–187.
41. Lee, D.; and Weitz, D. A. (2008) Double emulsion-templated nanoparticle colloidosomes with selective permeability, *Adv. Mater.*, **20**, pp. 3498–3503.
42. Duan, H.; Wang, D.; Sobal, N. S.; Giersig, M.; Kurth; D. G.; and Möhwald, H. (2005) Magnetic colloidosomes derived from nanoparticle interfacial self-assembly, *Nano Lett.*, **5**, pp. 949–952.
43. Mak, W. C.; Bai, J.; Chang, X. Y.; and Trau, D. (2009) Matrix-assisted colloidosome reverse-phase layer-by-layer encapsulating biomolecules in hydrogel microcapsules with extremely high efficiency and retention stability, *Langmuir*, **25**, pp. 769–775.
44. Noble, P. F.; Cayre, O. J.; Alargova, R. G.; Velev, O. D.; and Paunov, V. N. (2004) Fabrication of "hairy" colloidosomes with shells of polymeric microrods, *J. Am. Chem. Soc.*, **126**, pp. 8092–8093.
45. Cayre, O. J.; Hitchcock, J.; Manga, M. S.; Fincham, S.; Simoes, A.; Williams, R. A.; and Biggs, S. (2012) pH-responsive colloidosomes and their use for controlling release, *Soft Matter*, **8**, pp. 4717–4724.
46. Nguyen, T. H.; Tan, A.; Santos, L.; Ngo, D.; Edwards, G. A.; Porter, C. J. H.; Prestidge, C.; and Boyd, B. J. (2013) Silica-lipid hybrid (SLH) formulations enhance the oral bioavailability and efficacy of celecoxib: An *in vivo* evaluation, *J. Control. Release*, **167**, pp. 85–91.
47. Liu, H.; Chen, J.; Shen, Q.; Fu, W.; and Wu, W. (2010) Molecular insights on the cyclic peptide nanotube-mediated transportation of antitumor drug 5-fluorouracil, *Mol. Pharm.*, **7**, pp. 1985–1994.
48. Barnaby, S. N.; Fath, K. R.; Tsiola, A.; and Banerjee, I. A. (2012) Fabrication of ellagic acid incorporated self-assembled peptide microtubes and their applications, *Colloids Surf. B: Biointerfaces*, **95**, pp. 154–161.

49. Donath, E.; Sukhorukov, G. B.; Caruso, F.; and Möhwald, H. (1998) Novel hollow polymer shells by colloid-templated assembly of polyelectrolytes, *Angew. Chem. Int. Ed.*, **37**, pp. 2202–2205.
50. Gaponik, N.; Radtchenko, I. L.; Gerstenberger, M. R.; Fedutik, Y. A.; Sukhorukov, G. B.; and Rogach, A. L. (2003) Labeling of biocompatible polymer microcapsules with near-infrared emitting nanocrystals, *Nano Lett.*, **3**, pp. 369–372.
51. Gaponik, N.; Radtchenko, I. L.; Sukhorukov, G. B.; and Rogach, A. L. (2004) Luminescent polymer microcapsules addressable by a magnetic field, *Langmuir*, **20**, pp. 1449–1452.
52. Skirtach, A. G.; Antipov, A. A.; Shchukin, D. G.; and Sukhorukov, G. B. (2004) Remote activation of capsules containing Ag nanoparticles and IR dye by laser light, *Langmuir*, **20**, pp. 6988–6992.
53. Zebli, B.; Susha, A. S.; Sukhorukov, G. B.; and Parak, W. J. (2005) Magnetic targeting and cellular uptake of polymer microcapsules simultaneously functionalized with magnetic and luminescent nanocrystals, *Langmuir*, **21**, pp. 4242–4265.
54. Shen, H. -J.; Shi, H.; Ma, K.; Xie, M.; Tang, L. -L.; Shen, S.; Li, B.; Wang, X. -S.; and Jin, Y. (2013) Polyelectrolyte capsules packaging BSA gels for pH-controlled drug loading and release and their antitumor activity, *Acta Biomater.*, **9**, pp. 6123–6133.
55. Gao, Z. G.; Choi, H. G.; Shin, H. J.; Park, K. M.; Lim, S. J.; Hwang, K. J.; and Kim, C. K. (1998) Physicochemical characterization and evaluation of a microemulsion system for oral delivery of cyclosporin A, *Int. J. Pharm.*, **161**, pp. 75–86.
56. Wu, J.; Wei, W.; Wang, L. Y.; Su, Z. G.; and Ma, G. H. (2007) A thermosensitive hydrogel based on quaternized chitosan and poly(ethylene glycol) for nasal drug delivery system, *Biomaterials*, **28**, pp. 2220–2232.
57. Huynh, D. P.; Im, G. J.; Chae, S. Y.; Lee, K. C.; and Lee D. S. (2009) Controlled release of insulin from pH/temperature-sensitive injectable pentablock copolymer hydrogel, *J. Control. Release*, **137**, pp. 20–24.
58. Satarkar, N. S.; and Hilt, J. Z. (2008) Magnetic hydrogel nanocomposites for remote controlled pulsatile drug release, *J. Control. Release*, **130**, pp. 246–251.
59. Groot, K. D. (1980) Bioceramics consisting of calcium phosphate salts, *Biomaterials*, **1**, pp. 47–50.
60. Hench, L. L. (1991) Bioceramics: From concept to clinic, *J. Am. Ceram. Soc.*, **74**, pp. 1487–1510.

61. Xia, W.; and Chang, J. (2006) Well-ordered mesoporous bioactive glasses (MBG): A promising bioactive drug delivery system, *J. Control. Release*, **110**, pp. 522–530.
62. Simchi, A.; Tamjid, E.; Pishbin, F.; and Boccaccini, A. R. (2011) Recent progress in inorganic and composite coatings with bactericidal capability for orthopaedic applications, *Nanomed. Nanotechnol. Biol. Med.*, **7**, pp. 22–39.
63. Vallet-Regí, M.; Balas, F.; and Arcos, D. (2007) Mesoporous materials for drug delivery, *Angew. Chem. Int. Ed.*, **46**, pp. 7548–7558.
64. Betsiou, M.; Bantsis, G.; Zoi, I.; and Sikalidis, C. (2012) Adsorption and release of gemcitabine hydrochloride and oxaliplatin by hydroxyapatite, *Ceram. Int.*, **38**, pp. 2719–2724.
65. Colilla, M.; Manzano, M.; and Vallet-Regí, M. (2008) Recent advances in ceramic implants as drug delivery systems for biomedical applications, *Int. J. Nanomed.*, **3**, pp. 403–414.
66. Grimandi, G.; Weiss, P.; Millot, F.; and Daculsi, G. (1998) In vitro evaluation of a new injectable calcium phosphate material, *J. Biomed. Mater. Res.*, **39**, pp. 660–666.
67. Bigi, A.; Boanini, E.; and Rubini, K. (2004) Hydroxyapatite gels and nanocrystals prepared through a sol–gel process, *J. Solid State Chem.*, **177**, pp. 3092–3098.
68. Liu, J. B.; Ye, X. Y.; Wang, H.; Zhu, M. K.; Wang, B.; and Yan, H. (2003) The influence of pH and temperature on the morphology of hydroxyapatite synthesized by hydrothermal method, *Ceram. Int.*, **29**, pp. 629–633.
69. Cao, L.; Zhang, C.; and Huang, J. (2005) Synthesis of hydroxyapatite nanoparticles in ultrasonic precipitation, *Ceram. Int.*, **31**, pp. 1041–1044.
70. Donadel, K.; Laranjeira, M. C. M.; Goncalves, V. L.; Favere, V. T.; de Lima, J. C.; and Prates, L. H. (2005) Hydroxyapatites produced by wet-chemical methods, *J. Am. Ceram. Soc.*, **88**, pp. 2230–2235.
71. Stupp, S. I.; and Braun, P. V. (1997) Molecular manipulation of microstructures: Biomaterials, ceramics, and semiconductors, *Science*, **277**, pp. 1242–1248.
72. Zhang, S.; Greenfield, M. A.; Mata, A.; Palmer, L. C.; Bitton, R.; Mantei, J. R.; Aparicio, C.; de la Cruz, M. O.; and Stupp, S. I. (2010) A self-assembly pathway to aligned monodomain gels, *Nature Mater.* **9**, pp. 594–601.
73. Newcomb, C. J.; Bitton, R.; Velichko, Y. S.; Snead, M. L.; and Stupp, S. I. (2012) The role of nanoscale architecture in supramolecular

templating of biomimetic hydroxyapatite mineralization, *Small*, **8**, pp. 2195–2202.

74. Shanthi, P. M. S. L.; Mangalaraja, R. V.; Uthirakumar, A. P.; Velmathi, S.; Balasubramanian, T.; and Ashok, M. (2010) Synthesis and characterization of porous shell-like nano hydroxyapatite using Cetrimide as template, *J. Colloid Interface Sci.*, **350**, pp. 39–43.

75. Zhang, J.; Jiang, D.; Zhang, J.; Lin, Q.; and Huang, Z. (2010) Synthesis of organized hydroxyapatite (HA) using triton X-100, *Ceram. Int.*, **36**, pp. 2441–2447.

76. Wu, Y.; and Bose, S. (2005) Nanocrystalline hydroxyapatite: Micelle templated synthesis and characterization, *Langmuir*, **21**, pp. 3232–3234.

77. Verma, G.; Barick, K. C.; Manoj, N.; Sahu, A. K.; and Hassan, P. A. (2013) Rod-like micelle templated synthesis of porous hydroxyapatite, *Ceram. Int.*, **39**, pp. 8995–9002.

78. Lange, G. L.; and Donath, K. (1989) Interface between bone tissue and implants of solid hydroxyapatite or hydroxyapatite-coated titanium implants, *Biomaterials*, **10**, pp. 121–125.

79. Mavis, B.; and Tas, A. C. (2000) Dipcoating of calcium hydroxyapatite on Ti-6Al-4V substrates, *J. Am. Ceram. Soc.*, **83**, pp. 989–991.

80. Li, D.; Zhu, Y.; and Liang, Z. (2013) Alendronate functionalized mesoporous hydroxyapatite nanoparticles for drug, *Mater. Res. Bull.*, **48**, pp. 2201–2204.

81. Zhao, Y. F.; and Ma, J. (2009) Co-Synthesis and drug delivery properties of mesoporous hydroxyapatite-silica composites, *J. Nanosci. Nanotechnol.*, **9**, pp. 3720–3727.

82. Yang, P.; Quan, Z.; Li, C.; Kang, X.; Lian, H.; and Lin, J. (2008) Bioactive, luminescent and mesoporous europium-doped hydroxyapatite as a drug carrier, *Biomaterials*, **29**, pp. 4341–4347.

83. Xu, Q.; and Czernuszka, J. T. (2008) Controlled release of amoxicillin from hydroxyapatite-coated poly(lactic-co-glycolic acid) microspheres coated poly(lactic-co-glycolic acid) microspheres, *J. Control. Release*, **127**, pp. 146–153.

84. Palazzo, B.; Sidoti, M. C.; Roveri, N.; Tampieri, A; Sandri, M.; Bertolazzi, L.; Galbusera, F.; Dubini, G.; Vena, P.; and Contro, R. (2005) Controlled drug delivery from porous hydroxyapatite grafts: An experimental and theoretical approach, *Mater. Sci. Eng. C*, **25**, pp. 207–213.

85. Watanabea, K.; Nishiob, Y.; Makiurac, R.; Nakahirab, A.; and Kojimac, C. (2013) Paclitaxel-loaded hydroxyapatite/collagen hybrid gels as drug

delivery systems for metastatic cancer cells, *Int. J. Pharma.*, **446**, pp. 81– 86.

86. Yang, Y. -H.; Liu, C. -H.; Liang, Y. -H.; Lin, F. -H.; and Wu, K. C. -W. (2013) Hollow mesoporous hydroxyapatite nanoparticles (hmHANPs) with enhanced drug loading and pH-responsive release properties for intracellular drug delivery, *J. Mater. Chem. B*, **1**, pp. 2447–2450.

Chapter 3

New Trends in Bioactive Glasses: The Importance of Mesostructure

Sevi Murugavel and Chitra Vaid
*Department of Physics and Astrophysics, University of Delhi,
Delhi-110007, India*
murug@physics.du.ac.in

Bioactive glasses are the most promising bone replacement/ regenerative materials because they bond to natural bone, are degradable, and stimulate new bone growth by the action of their dissolution products on cell. However, they lag behind other bioactive ceramics in terms of commercial success. In this chapter, we review the main developments in the field of bioactive glass and its variants, covering the importance of control of hierarchical structure, synthesis, processing, and cellular response in the quest for new regenerative synthetic bone grafts. Among the various strategies for the development of bioactive glasses, the sol-gel-derived mesoporous bioactive glasses (MBGs) can be foamed to produce interconnected macrospores suitable for the tissue ingrowth, specifically cell migration and vascularization and cell penetration. The combination of excellent textural parameters and

Trends in Biomaterials
Edited by G. P. Kothiyal and A. Srinivasan
Copyright © 2016 Pan Stanford Publishing Pte. Ltd.
ISBN 978-981-4613-98-9 (Hardcover), 978-981-4613-99-6 (eBook)
www.panstanford.com

hence an enhanced *in vitro* bioactivity make the wormhole-like MBGs promising for applications in clinical orthopedics, controlled drug delivery, tissue engineering, etc. This chapter also focuses on the techniques that have been developed for characterizing the hierarchical structure of sol–gel glasses and hybrids, from atomic-scale disordered structures, through the covalent network between the components in hybrids and nanoporosity, to quantify the open macroporous networks of scaffolds. We also provide to the readers the recent advancement in this field from Hench's 45S5 Bioglass® to new hybrid materials that have tunable properties and superior degradation rates.

3.1 Historical Background

Tissue engineering has emerged as the most common approach for the regeneration and repair of tissues and organs lost or damaged as a result of trauma, injury, disease, and aging. It has the potential to overcome the shortage of living tissues and organs available for transplantation. In the past few decades, tissues such as skin, bone, and cartilage have been successfully regenerated. It is important to mention that more than 2.2 million bone graft operations are carried out worldwide every year to repair bone-related defects in orthopedics and dentistry [1, 2]. In this context, autografts are the most preferred choice for the treatment of bone defects, but their limited supply and donor-site morbidity lead to serious problems. Being a natural tissue, the successful remodeling of bone strongly depends on various factors such as blood vessel cells, progenitor cells, and growth factors. Alternatively, bone allografts have been developed, but they have many drawbacks such as they are expensive, have the risk of disease transmission, and possess poor mechanical properties [3–5]. Hence, synthetic biomaterials are ideal for bone substitutes. The commercial success of synthetic biomaterials is limited since they do not meet all the requirements of natural bone that for an autologous bone. More commonly, implants for bone replacement or fracture repair in load-bearing applications are made from metallic alloys for mechanical support; typically Ti6Al4V or Co–Cr alloys are used for joint/knee replacements [6, 7]. Although these metallic alloys are stronger and stiffer than human bone, they promote resorption by shielding the surrounding skeleton from its normal levels and ultimately the implant becomes movable over a period of time. Subsequently, more products are designed to be bioactive such that they stimulate a specific biological response at the material surface, which leads to the formation of a bond between the tissue and the material. A typical example is synthetic hydroxyapatite (HAp), which has a chemical composition similar to that of the bone mineral. However, the slow biodegradability and low mechanical strength of HAp ceramics restrict their use in tissue engineering applications [8–10].

Bioactive glasses are one of the most attractive synthetic bone replacement materials available since they are potentially more bioactive than pure calcium-phosphate-derived materials. In 1969, Hench and his coworkers discovered that bone could chemically bond to glasses having certain compositions. This group of glasses is referred to as bioactive glasses or bioglasses [11, 12]. An important characteristic of these glasses is the formation of a hydroxycarbonate apatite (HCA) layer on their surface in aqueous solutions. Bioactive glasses can be produced in a wide range of forms, and they serve various functions in the body. The bioactive glass 45S5 Bioglass (46.1% SiO_2, 24.4% Na_2O, 26.9% CaO, and 6% P_2O_5, in mol%) was the first material produced by Hench in 1971 [11, 12]. After Hench's discovery, various kinds of glasses have been synthesized by numerous research groups. However, no other bioactive glass composition has been found to have superior biological response over the original 45S5 glass composition. Due to the differences in composition, structure, and constituent phases, the bone-bonding properties of these materials differ from each composition. The relative bioactivity among different materials can be evaluated by measuring the rate of bone formation on the surface of the material.

The 45S5 Bioglass, which was prepared by the conventional melt-derived method at the earlier stages, is most widely studied and used in clinical applications. The abbreviation indicates that it contains 45% SiO_2 by weight and that the molar ratio between Ca and P is 5:1. Three key compositional features of the SiO_2–Na_2O–CaO–P_2O_5 bioactive glasses distinguish them from traditional soda-lime–silica glasses: (1) less than 60 mol% SiO_2, (2) high Na_2O and CaO content, and (3) high CaO/P_2O_5 ratio. These are the salient features that make the surface of bioactive glasses highly reactive when exposed to an aqueous medium.

The 45S5 bioactive glass finds application in middle ear prostheses to restore the ossicular chain and treat conductive hearing loss. It is also used as oral implants to preserve the alveolar ridge from bone resorption that follows tooth extraction and as a coating for artificial dental roots that could be used as self-standing dental implants. In the 1990s, bioactive glass composites had become the standard for repairing and replacing bones in the middle ear. However, they also suffered from various limitations [13] such as very high temperature for melting and lack of micro-/mesoporous structure leading to low specific surface area. Therefore, the bioactivity of these glasses depended mainly on silica content. Besides silicate glasses, phosphate glasses were also studied for bone regeneration application. It has been suggested that unlike silica-based glasses, phosphate-based glasses, which are similar to the inorganic part of the bone, have the unique property that their degradation rate can be adjusted by adding various oxides according to end application [14]. Although phosphate bioceramics and HAp ceramics have been widely used for bone replacement and regeneration, HAp ceramics seem to have poor *in vitro* bioactivity and degradation compared to bioactive glasses [15]. Therefore, sol–gel-derived bioactive glasses came in as an effort to overcome the limitations of melt-quenched bioactive glasses. Due to greater surface area and inherent porosity, they showed a wider range of bioactive composition and exhibited a higher rate of bone bonding, degradation, and resorption properties. However, these sol–gel-prepared glasses also suffered from some limitations such as nonuniform micropore distribution and inadequate drug loading and release [16]. Thus, to overcome the limitations of conventional bioactive glasses, new generations of bioactive glasses were designed and developed by combining efficient drug delivery and excellent bioactivity. A new concept was developed by Vallet-Regi [17], and Yu and coworkers [18] simultaneously reported that these glasses could be prepared via the combination of the sol–gel method and the supramolecular chemistry of surfactants. They were able to synthesize MBGs based on $CaO-SiO_2-P_2O_5$ composition and have highly ordered mesoporous channel structure and narrow pore size distribution. Compared to the conventionally prepared sol–gel bioactive glasses, MBGs possess more optimal surface area and pore volume, greater ability to induce *in vitro* apatite mineralization in simulated body fluid (SBF), and excellent cytocompatibility [19].

3.1.1 Scaffolds for Bone Tissue Engineering

Tissue engineering tries to promote the regeneration ability of the host tissue through a designed scaffold integrated with cells and signaling the molecules. The tissue engineering approach has been investigated in the recent years to tackle the growing clinical demand for engineered bone tissue due to increasing population. Following this strategy, a three-dimensional structure, termed "scaffold" or template, is often designed using materials that safely dissolve once performing their function. For scaffold fabrication, a suitable artificial or natural material that exhibits high porosity and pore interconnectivity is required. The scaffolds, apart from providing passive structural support to the bone cells, should favorably affect bone formation by stimulating osteoblastic cell proliferation and differentiation. Bioactive glasses have high potential as scaffold materials, which make them suitable for bone regeneration as they stimulate bone cells to produce new bone, are biodegradable, and bond to the bone (i.e., they are osteoconductive as well as osteoproductive). Hence, the ideal scaffold for bone tissue engineering should possess the following requirements: (1) ability to promote cell adhesion and proliferation; (2) an appropriate mechanical property compatible with the host tissue; (3) a three-dimensional porous architecture that acts as a template for three-dimensional bone growth; (4) good biodegradability; and (5) capability of being processed economically into relevant shapes and sterilized for clinical use and commercialization. More commonly, scaffolds for tissue engineering applications are made from biodegradable synthetic or natural polymeric materials [20]. However, the use of biodegradable polymers as scaffolds is challenging because of their poor mechanical property. To improve mechanical strength, an effort has been made to fabricate a biocompatible inorganic phase reinforced with biodegradable polymers such as HAp.

3.1.2 Bioactive Glass Scaffolds

Bioactive glasses have been synthesized by either the melt-quenching or the sol–gel process over the decades. Perhaps the greatest advantage of opting for sol–gel synthesis is that this technique has led to the formation of porous scaffolds with interconnected macropores inherently suitable for tissue engineering application.

Since the invention of bioactive glass in 1969, no successful porous scaffold has been synthesized, since the sintering process employed to produce porous structure leads to crystallization of bioglass. Therefore, a sintered bioglass scaffold is actually a glass-ceramic or ceramic scaffold. In bone tissue engineering applications, cells play a vital role. The dimensions of the cells are in the range of 10–200 nm, whereas the pore dimension of mesoporous materials falls in the range of 20–50 nm. Hence, mesopores are too small to allow cell uptake; therefore, the designing of hierarchically three-dimensional scaffolds with interconnected macroporosity is a prerequisite for bone tissue engineering. Such macroporosity will enhance bone cell penetration, adherence, growth, and proliferation, leading to bone tissue ingrowth on implantation.

Three-dimensional macroporous scaffolds based on simple binary SiO_2–CaO and ternary SiO_2–CaO–P_2O_5 compositions have been developed by several research groups [21]. Although macroporous scaffolds exhibit better *in vitro* bioactive behavior than non-three-dimensional structured glasses, an increase in macropore size slowed down the glass degradation and apatite formation process. Foaming of sol–gel glasses produces scaffolds mimicking cancellous bone macrostructure [22]. Foamed bioactive scaffolds show higher osteoblast proliferation as well as collagen secretion. Compressive mechanical properties can be improved by optimizing nanoporosity of these foams, although toughness and strength are the prevailing problems. Furthermore, giant macro-mesoporous bioactive three-dimensional glass scaffolds have been synthesized by combination techniques resulting in scaffolds with multiple values of porosity. SBF *in vitro* tests reveal the preservation of bioactive behavior and formation of apatite layer after 24 h of assay. The scaffolds' structures can be organically modified due to the presence of silanol groups in the external surface. The choice of functional groups enables the chemical grafting between material surface and osteoinductive agents that signal appropriate cellular functions where needed [23].

3.1.3 Mesoporous Bioactive Glass Scaffolds

Silica-based mesoporous materials (SMMs) such as SBA-15, MCM-48, and MCM-41 are characterized by high surface area, pore volume, and pore size, with narrow pore size distribution. In addition, the SMMs possess high surface silanol density, which confers the necessary requisites to behave as bioactive materials. Furthermore, from *in vitro* assays performed on SBA-15, MCM-48, and MCM-41, it has been observed that high surface area and high silanol density are not enough to promote bioactive behavior [24]. As a result, various strategies were employed [25] to accelerate the bioactive response of SMMs. In practice, a significant advancement was achieved in this field when the sol–gel synthesis of MBGs was developed with SiO_2, CaO, Na_2O, and P_2O_5 as the main components, with outstanding textural properties and ordered pore arrangement [26]. By retaining the mesostructure, glasses showed exceptional HCA growth in SBF with shorter period of time.

The *in vitro* bioactive behavior of MBGs is much faster and intense than that of conventional sol–gel or melt-derived glasses. The ion-dissolution kinetics in these glasses is very intense, resulting in massive and fast amorphous calcium phosphate formation onto the surface of MBGs. Moreover, it was found that mineral maturation in these materials is equivalent to vertebrates. Most of the MBG compositions were able to develop crystalline HAp very fast, and hence various research groups are aiming toward the preparation of scaffolds for bone tissue engineering or *in situ* implantation. Due to their unique properties, MBGs proved to be an excellent candidate for biomedical applications such as drug delivery systems [27] and bone tissue regeneration [28]. Due to physicochemical characteristics exhibited by MBGs, the Hench mechanism of HCA formation has been reformulated. The mechanism formulated for the formation of HCA onto the MBG surface is similar to that proposed by Hench *et al.* for conventional bioglass.

However, there are some important differences that lead to an accurate biomimetism in MBG with respect to natural bone [29]. Such differences are summarized in Fig. 3.1. The steps followed in Fig. 3.1 are as follows:

Stage 1: There is rapid exchange of alkali–alkaline–earth ions with H^+ from the surrounding fluid medium. In the case of MBG, high accessible porosity, surface area, and material reactivity accelerate the surface process, which permits a more intense ionic exchange and higher H^+ incorporation compared to conventional bioglass.

Stage 2: In this stage, silanol (Si–OH) groups form at the glass surface. In the case of MBGs, the density of silanol groups is higher than in the case of conventional bioglass because of high surface area and larger H^+ incorporation.

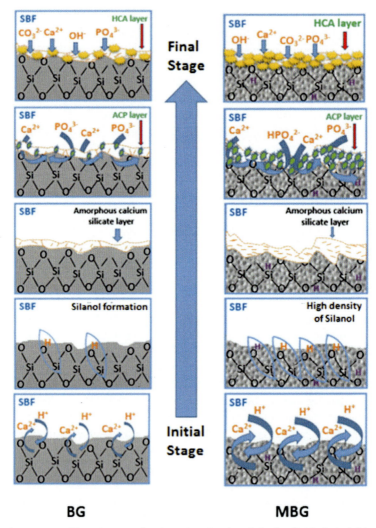

Figure 3.1 Bioactive mechanism in simulated body fluid (mimicking human plasma) of conventional bioglass (Hench, 1970) vs. MBG (Vallet-Regi, 2008).

Stage 3: It is observed that the polycondensation of silanol groups forms a hydrated silica gel, followed by a highly protonated silica-rich layer with depleted cations.

Stage 4: The Ca^{2+} and PO_4^{3-} groups migrate to the surface through the silica-rich layer and from the surrounding fluid forming a CaO–P_2O_5-

rich layer on the top of the silica-rich layer, followed by the growth of amorphous CaP (ACP). In the case of MBGs, the ACP precipitation is higher compared to that in conventional sol–gel glasses.

Stage 5: The crystallization of ACP leads to the formation of HCA by the incorporation of Ca^{2+} and HPO_4^{2-} ions. Thus, the overall time period for HCA formation varies in both the glasses.

Therefore, the texture–property correlations play an important role in the formation of HCA layer in MBGs, which is much different from that proposed for conventional bioactive glasses, due to high surface area, pore size, and pore volume.

3.2 Fabrication of Bioactive Glass and Glass-Ceramic

3.2.1 Melt-Derived Bioactive Glasses

Bioactive glasses are typically prepared from high-purity raw materials since the quality strongly influences the end product. Accurately weighed amount of raw materials are melted in a platinum or platinum–rhodium crucible, and then it is homogenized as required according to the composition. The melting temperatures can range from 1200 to 1450°C depending on the chemical composition. The melt is cast in stainless steel molds, which do not contaminate or adhere to the glass or glass surface. For typical synthesis of 45S5 Bioglass®, initially the reagents Na_2CO_3 and $CaCO_3$ were dried at 473 K to remove any traces of water and adsorbed gases. Then the stoichiometric amounts of the starting materials were thoroughly ground in an agate mortar for 50 min or using the planetary ball mill. The mixture was then placed in a platinum crucible and melted in an electric furnace for 5 h at 1773 K. After complete homogenization, the melt was poured into preheated stainless steel molds [30]. However, it has been found that the bioactivity of melt-derived glasses is highly composition dependent. Compositions with 60% and higher SiO_2 content do not develop HAp layer even after several weeks in stimulated body fluid and bond to neither bone nor soft tissues. Thus, to increase the range of compositions showing bioactive behavior, focus shifted to sol–gel synthesis process in addition to the superior textural properties.

3.2.2 Sol–Gel-Derived Bioactive Glasses

Sol–gel glasses were obtained by the hydrolysis and polycondensation of organic precursors in aqueous environment with an acidic or basic catalyst. More commonly, ethanol was added to the starting solution to help mixing organic and aqueous phases. The solution was continuously stirred for a sufficiently long time, and organic precursors were hydrolyzed. In parallel with Si–OC$_2$H$_5$ bond hydrolysis, a process of recondensation occurs between the silanol groups. A polymerized network is obtained, which finally becomes a gel. After gelation, the sample is aged usually at room temperature and then dried at 120–150°C to eliminate excess water and the alcohol obtained from the hydrolysis. Subsequently, the dried gels can then be stabilized at higher temperatures [31]. For preparing MBGs, a modified sol–gel method has been developed with polymer or surfactant used as a template or structure-directing agent (SDA) to generate mesopore structure in the resulting glass matrix. The synthesis of MBGs involved the addition of surfactant as SDA to the conventional sol–gel glass synthesis using the evaporation-induced self-assembly (EISA) process, which plays a keystone for successful preparation of this new generation of bioactive glasses.

3.2.2.1 Supramolecular chemistry and sol–gel process

A new generation of mesoporous biologically active glasses has been prepared by the incorporation of supramolecular chemistry to the sol–gel process [31]. This new generation of glasses exhibits outstanding porosity and surface area values and is capable of hosting active agents that contribute to the tissue-healing processes. In the course of fabrication, to obtain successful structures, the incorporation of SDA (mainly block copolymers) is essential. Under desired synthesis conditions, these molecules self-organize into micelles and link the hydrolyzed silica precursors through the hydrophilic component and self-assembly to form mesophase. The mesophase ordering depends on several factors such as surfactant chemistry (ionic, nonionic, and polymeric), organic–inorganic phase volume ratio, surfactant concentration, temperature, and pH. The incorporation of nonionic surfactants such as triblock copolymers, for instance (EO)–(PEO)–(EO), as SDAs opened new possibilities for obtaining MBGs. These SDAs combined with the EISA process were the keystone for the successful preparation of this new generation of MBGs, and we illustrate the typical EISA process in Fig. 3.2.

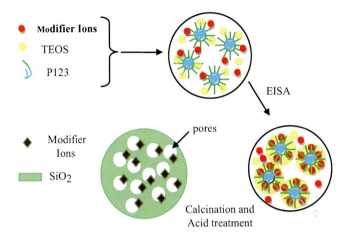

Figure 3.2 Schematic representation of EISA process.

3.2.3 Flame Spray Synthesis

The gas phase synthesis or flame spray synthesis method is used for the fabrication of nanoparticles at temperatures above 1000°C. This method was originally used to manufacture carbon black and subsequently extended to the synthesis of various classes of materials, including nano-bioglass. The flame spray technique is a gas-phase-based method that uses metal–organic precursors to produce nanoparticles at temperatures above 1000°C [32]. The liquid precursor is dispersed by oxygen over a nozzle thereby forming a spray, which is ignited. As the spray is burning, the organic constituents of the liquid precursor completely combust mainly to water and carbon dioxide and metal oxide constituents to form the nanoparticles. This process enables the production of numerous nanoparticulate mixed-oxides with high chemical homogeneity; by using this technique, the preparation of nanoparticles of size about 20–80 nm and with specific surface area of 70–90 m^2/g in different glass compositions has become possible. Although the flame synthesis procedure is promising, the production environment employed in this protocol becomes highly inconvenient, in addition to poor textural parameters.

3.2.4 Freeze-Drying Technique

The freeze-drying technique or thermally induced phase separation method is used to fabricate porous composite bioactive glass scaffold structures. In this method, phase separation is induced by lowering the temperature of the suspension of polymer and ceramic mixture into the interstitial spaces. The frozen mixture is then lyophilized using a freeze dryer, in which the ice solvent evaporates. After drying, the pores constructs are sintered to remove the fine pores between the particles in the walls of the macropores, which leads to improvement in the mechanical strength. The directional freezing of the suspensions leads to the growth of the ice in a favored orientation, resulting in the formation of porous scaffolds with uniform microstructure. Therefore, the main processing steps involved are mixing of key ingredients such as powder, solvent, and functional and processing additives followed by desecration and casting. The solidification is followed by sublimation and sintering as per the requirement of ending up with the sample required [33].

3.3 Silica-Based Sol–Gel Bioactive Glass Composites

Bioactive glass composites are materials with new properties needful for several medical and technological applications. Composite materials show unique characteristic features as they combine properties of traditional bioactive glass with new materials such as polymer or other organic materials at nanoscopic scale. More commonly, composite materials exhibit properties of both organic and inorganic counterparts present in the materials, due to which they show important advancement in the field of hybrid materials with enhanced properties than the traditionally prepared glasses. These materials help to overcome the main drawback of silicate-based glasses: inherent brittleness. Due to their brittleness, these glasses are not applied in load-bearing applications and to fill small bones. Thus, composite materials are synthesized to achieve better mechanical strength along with bone-bonding ability. Composite materials provide hardness, strength, stability as well as chemical reactivity and hydrophobicity, making them technologically and biologically important materials for biomedical applications. Interfacial interactions between organic and inorganic precursors in composite materials play a vital role in ensuring enhanced mechanical strength for load-bearing applications. Therefore, composite materials are synthesized using components in such a way that more and more hydrogen interactions are possible.

Silicate-based glasses provide better bioactivity to the organic component in composite materials. Composite materials were first prepared by poly(methyl methacrylate)-silicate system for dental restoration and bone application [34]. This system showed weak *in vitro* behavior but better cell attachment, proliferation, and differentiation when placed in mouse calvarial osteoblast cell culture. Polyethylene glycol-SiO_2 composite materials also exhibit *in vitro* bioactivity to a certain extent. The widely studied bioactive glass composite is synthesized using poly dimethylsiloxane as precursor [35]. These composites developed apatite layer within 1 day of SBF treatment, and it was observed that the presence of silanol groups played a crucial role in the bioactive behavior. Other bioactive glass composite materials include 3-methacryloxyproply-trimethoxysilane and hydroxy(ethyl methacrylate), which promote better ion exchange, and gelatine-silicate composites, which have potential for bone tissue regeneration. Recently, star-gels have been synthesized with the combination of unique properties such as mechanical strength and biocompatibility, making them excellent candidates for bone regeneration in medium and large bone defects [36].

3.4 Hybrid Sol–Gel Bioactive Glasses

Hybrid materials are organic–inorganic constituents of interpenetrating network components that interact at the nanoscale level [37]. Hybrid materials are different from nanocomposites and have distinguishable components. On the other hand, in the case of hybrid materials, a number of components present become indistinguishable above the nanoscale level. More importantly, the synthesis of hybrid materials is complex, and several issues must be overcome before hybrid materials are exploited in tissue engineering applications. In general, hybrid materials are synthesized by introducing the polymer at the earlier stage of the sol–gel process, e.g. after the hydrolysis of tetraethyl orthosilicate (TEOS), which stimulates the inorganic network formed around the polymer molecules, resulting in atomic-scale interactions. In general, most hybrid systems are aged and dried below 100°C. An important hypothesis is that the fine-scale interactions between the organic and inorganic chains lead to the material behaving as a single phase, resulting in controlled congruent degradation and the potential candidate for the tailored mechanical properties [37]. The fine-scale dispersion of the different components means that the cells are likely to attach to the hybrid surface as though it is a single component, rather than bioactive particles are dispersed in a polymer matrix. The novelty is that a bioactive hybrid would have different bioactivity similar to that of a bioactive glass, but they have enhanced mechanical property.

In hybrid synthesis, the chain-like structure of inorganic constituents can entangle with the polymers since they are added to the sol at the earlier stages of the condensation process. More commonly, under the acidic conditions, the TEOS hydrolysis and then condensation continue forming coalesce and then coordinate to form a gel. Additionally, it is interesting to note that the control of pH is important during the entire process, where it can affect the functionalization of the polymer and the gelation of the inorganic network structure. Hence, the pH can also cause degradation of polymers during the hybrid synthesis.

3.5 *In Vitro* Response of Bioactive Mesoporous Glasses

The importance of analyzing bioactivity by *in vitro* prior to *in vivo* analysis is quite clear. More commonly, *in vivo* studies require animal sacrifice, are more costly and less easily reproducible as well as involve ethical issues. For these reasons, before the use of these materials in *in vivo* applications, *in vitro* tests are highly necessary. The choice of the solution used to simulate *in vitro* reactions occurring on the surface of the biomaterial is very important: simple solutions that mimic only the inorganic composition of human body fluids can be used or more complex solutions that contain biological moieties such as proteins. Moreover, solutions containing cells can be employed, thus increasing both the similarity to real body fluids and the complexity of the test.

The kinetics of dissolution of the glass scaffold and conversion to HCA by *in vitro* process have been studied by the immersion of glass in SBF at 37°C and measuring the weight loss of the glass as a function of time as well as by monitoring the pH of the solution [38–40]. Additionally, the conversion product has been characterized by structural, chemical, morphological, and nanomechanical probes and confirms the obtained product phase. The systematic investigations on various glass systems indicate that the kinetics of conversion to HCA strongly depends on the particular glass composition [30, 31]. More commonly, it has been observed that the reaction is pseudomorphic, starting at the surface of the glass and moving inward. It has been observed in borate glasses that the conversion is controlled by the reaction at the interface and the kinetics can be described by a three-dimensional contracting sphere model. In the case of silicate glasses, the reaction is controlled initially by the interface and later by the diffusion of ions to the reaction interface. In general, it is accepted that if the reaction time is sufficiently long for the product to crystallize, where the X-ray diffraction (XRD) pattern of the converted product shows peaks that correspond to the reference HCA phase. However, studies on ternary and quaternary silicate glasses show the amorphous nature of the converted product [31]. Hence, the biomineralization of bioactive glasses strongly depends on the glass composition, and it is not yet completely understood. Additionally, the Ca/P ratio of the converted material often varies from the value of stoichiometric HCA, where it varies from the surface of the reacted glass to the interior [41, 42].

In vitro bioactive behavior of MBGs is much faster and intense compared to conventional sol–gel bioactive glasses due to much higher surface area and porosity. Mesoporous silica bioactive glasses have been studied *in vitro* as well as *in vivo* over the decades. It has been observed that the slow kinetics of HCA formation during *in vitro* process, however, proved to be efficient in repairing cavity defects under *in vivo* conditions. The recent results show that SBA-15 loaded with oseostatin promotes new bone growth in osteoprotic animals. *In vitro* studies on binary mesoporous glass system such as SiO_2–CaO and SiO_2–Na_2O revealed that in mesoporous binary systems, apatite crystallization takes place within 72 h of SBF immersion compared to corresponding melt-derived counterparts which took 240 h for crystallization [28]. *In vitro* studies on multicomponent SiO_2–CaO–P_2O_5 system reveal that both composition and texture play a very important role in determining the chemical reactivity of the system when placed in SBF.

A mesoporous multicomponent system offers outstanding *in vitro* behavior. It has been observed that the *in vitro* bioactivity of these glasses is highly dependent on the structural properties. The sol-gel glasses of the multicomponent system have been found to exhibit faster HCA formation both *in vitro* and *in vivo*, in addition to the extensive ingrowth. The *in vivo* studies on 77S (77SiO_2–19CaO–4P_2O_5), 58S (58SiO_2–38CaO–4P_2O_5), and 45S5 melt-derived composition show similar cellular response upon implantation [28]. Comparative studies for *in vitro* bioactivity on mesoporous glasses in quaternary (SiO_2–CaO–Na_2O–P_2O_5) and ternary (SiO_2–CaO–Na_2O and SiO_2–Na_2O–P_2O_5) glasses show that the joint presence of calcium and phosphorous in the quaternary silicate glasses accelerated the growth rate of HCA formation than the ternary glass system, where similar textural properties yield poor HCA formation [31]. Moreover, *in vitro* and *in vivo* studies on ion-substituted bioactive glasses reveal the role of different ion on bone metabolism and on angiogenisis, growth, and mineralization of bone tissue. It has been proved that silicon is effective for the formation and calcification of bone tissues. Additionally, studies show that silicon supplementation on calcium-deficient rats caused increased bone mineral density [32].

The chemical constituents of bioactive glass have a significant effect on its ability to support the proliferation and function of cells in the *in vitro* process. The conventional silicate bioactive glasses in the form of disks or porous scaffolds are found to support the proliferation and differentiation function of osteoblastic cells such as murine MC3T3-E1 and MLO-A5 cells [43, 44]. In contrast, borate glasses have shown a pronounced lower ability to support cell proliferation and function. These features are found to be due to the faster degradation rate of borate glass and to the toxicity of the boron released into the culture medium. It has been shown that borate glasses lead to higher pH value of the culture medium.

3.6 *In Vivo* Response of Bioactive Mesoporous Glasses

A number of studies have shown the *in vivo* behavior of 45S5 bioactive glass where enhanced bone formation was observed upon implantation in animal body [45–47]. More interestingly, recent studies have shown the effect of bioactive glass composition on its ability to support tissue ingrowth by *in vivo* methods. Fu *et al.* have studied the implantation of bioactive scaffolds with the similar "trabecular" microstructure, but with different compositions subcutaneously in the dorsum of rats, and evaluated the microstructure and histology of the scaffolds after six weeks [48]. An *in vivo* study by Wang and coworkers on $30CaO \cdot 70SiO_2$ (70S30C) bioactive glass shows soft tissue infiltration within the porous scaffold in a rabbit subcutaneous model. However, the main drawbacks of these studies are poor histological confirmation and reliability [49]. Moreover, the ability of the scaffold to permit soft tissue ingrowth within the pores does not well predict the bone-forming potential of the material. More recently, Swati *et al.* have reported the *in vivo* behavior of preconditioned 70S30C bioactive glasses prepared by the sol–gel process and revealed that they are very effective in regenerating a defective rat tibia within 11 weeks, showing robust bone ingrowth throughout the pore network and degradation of the scaffold structure [50]. In this case, the bioactive glass composition is not optimal for scaffold design since scaffolds were not pre-treated or were only shortly wetted, demonstrating minimal bone ingrowth. Furthermore, they have compared the bone ingrowth of 70S30C scaffolds with that of commercial synthetic bone grafts (NovoBone® and Actifuse®). It has been found that poor bone ingrowth was found for both the dry and wetted sol–gel foams, related with the rapid increase of pH within the scaffolds, and they were quantified through histology and novel micro-CT image analysis.

3.7 Local Structure of MBGs and Glass-Ceramic and its Relation with Dissolution Kinetics

The bioactive properties of these glasses such as degradation rate, formation of HCA layer, and mineralization behavior highly depend on the local atomic structure. Hence, an understanding of the structure of bioactive glass and designing newer compositions for biomedical applications are important. To understand the various aspects of HCA layer formation in different bioactive glasses, diverse experimental techniques such as XRD, Fourier transform infrared (FTIR), nuclear magnetic resonance (NMR) spectroscopy, and different microscopic analyses have been explored [51, 52]. These experimental techniques tend to be very useful in investigating the surface reactions involved in Hench mechanism, starting from leaching of ions to HCA formation. More recently, the modifications of the surface composition upon SBF immersion as well as the deposition of Ca–P layer and its consequent crystallization to HCA layer at the MBG surfaces have been investigated [53]. Additionally, combined NMR and high-resolution transmission electron microscopy (HRTEM) investigations have been carried out to probe the mechanism of apatite mineralization, and these investigations reveal that the surface reactions observed in MBGs are in contrast with Hench mechanism. Hence, the Hench mechanism of apatite formation is more applicable for traditional melt-derived bioactive silicate glasses (MDBG) due to the local atomic structure and poor textural nature [54].

Based on the obtained experimental results, different conclusions have been made to correlate the local structure and formation of the HCA layer. However, more details of bioactive glasses, such as inter-correlation between chemical composition and HCA layer formation and surface reactions involved in it, are under debate. In this context, computer simulation techniques have been exploited, which predict the bioactive behavior of these glasses and their local atomic structure [55]. In particular, the work done by Tilocca and his coworkers by molecular dynamics simulation study offers fundamental understanding of the structure of the complex nature of 45S5 glass and a few other glass compositions [56]. Recently, the local environments of the network forming species at the atomic level and the interrelation with the HCA formation [57] have been shown by MAS-NMR investigations on quaternary bioactive silicate glasses. Figure 3.3 shows the ^{29}Si MAS-NMR spectra of three different bioactive glass and glass-ceramic samples (53S30C–53.4%SiO_2–30CaO–14Na_2O–2.6P_2O_5, 63S20C–63.4%SiO_2–20CaO–14Na_2O–2.6P_2O_5, 73S10C–73.4%SiO_2–10CaO–14Na_2O–2.6P_2O_5) identified to obtain the local structural speciation of various Q^n species. Here we adopt the Q^n notation to denote silicon and phosphorus atoms bonded to n bridging oxygen (BO) and (4-n) non-bridging oxygen (NBO) atoms.

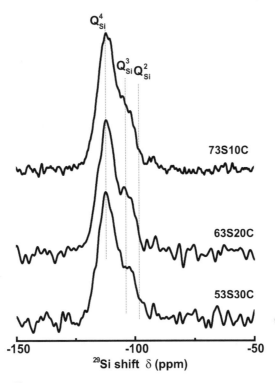

Figure 3.3 ^{29}Si MAS-NMR spectra of representative pristine sol–gel bioactive glass and glass-ceramic.

All the pristine samples are dominated by a broad band centered at −112 ppm assigned to Q_{Si}^4 species, which form the three-dimensional network structures. The spectral assignments are less obvious because of the competing roles of network modifier and H$^+$ for charge compensating the NBOs present. In bulk alkali and alkaline-earth modified silicate glasses, the chemical shift value for Q_{Si}^2 (Na) is assigned at −76 ppm and that for Q_{Si}^2 (Ca) is assigned at −83 ppm [58]. However, we do not observe any spectral features at these chemical shift values in the obtained raw spectra of present glass compositions. To evaluate more qualitatively the variation of individual Q^n species with CaO content, we have followed the deconvolution procedure and identified the species associated within the glass and glass-ceramics. On the deconvoluted spectra, we have found two major components corresponding the Q_{Si}^4 and Q_{Si}^3 species and a minor amount of Q_{Si}^2 species due to the depolymerization of silica network (Fig. 3.4). The MBGs that involve Q_{Si}^3 and Q_{Si}^2 structural units are more commonly associated with $Si(OSi)_3OH$ and $Si(SiO)_2(OH)_2$ groups due to the presence of silanol (Si–OH) groups [59]. It has been observed more qualitatively that the relative proportion of Q_{Si}^3 band decreases in relation to reduction in the modifier concentration. Table 3.1 lists the relative populations (expressed as percentages) of Q_{Si}^n species obtained from the deconvolution of all the spectra. The variation of silica Q^n species in these glasses and glass-ceramic evidences the effect of network modifiers on the silicate network.

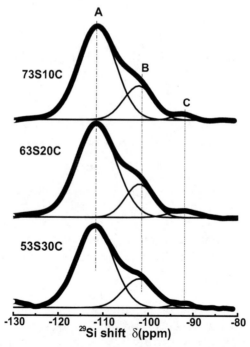

Figure 3.4 ^{29}Si spectra for pristine glasses and glass-ceramic after deconvolution into various Q species for silica. (Open circle represents the obtained spectra, and solid line represents deconvoluted spectra).

Table 3.1 Relative populations (expressed as percentages) of Q_{Si}^n species obtained after deconvolution of the obtained spectra

Sample	Q_{Si}^4 (A) δ (ppm)	Area (%)	Q_{Si}^3 (B) δ (ppm)	Area (%)	Q_{Si}^2 (C) δ (ppm)	Area (%)
73S10C	−111.58	78	−102.3	20	−92.6	2
63S20C	−111.60	73	−102.2	23	−92.3	4
53S30C	−111.8	68	−101.9	25	−92.4	7

Fig. 3.5 shows the ^{31}P MAS-NMR spectra of pristine 73S10C, 63S20C, and 53S30C samples with distinguishable features. To get detailed insights on the type of Q_p^n species present in the glass and glass-ceramic samples, the ^{31}P NMR spectra are deconvoluted and fitted to Gaussian lines as shown in Fig. 3.6. The chemical shift values and the percentage of population (peak area) of individual species after deconvolution of the ^{31}P spectra are summarized in Table 3.2. Clearly, the ^{31}P NMR spectra of the 73S10C glass sample exhibit four resonance peaks at −3.1, −1.4, 0, and 1.2 ppm as shown in Fig. 3.6. The observed resonance peak around −3.1 ppm lies between the ^{31}P chemical shifts of $Ca_2P_2O_7$ (−7 ppm) and $Na_4P_2O_7$ (2.9 ppm) suggesting that phosphorous exists as pyrophosphate Q_p^1 complex with mixed sodium and calcium environment [60]. The chemical shifts at −1.4 ppm and 1.2 ppm could be assigned to dicalcium phosphate dehydrate (monetite, $CaHPO_4$) and dicalcium phosphate dihydrate (brushite, $CaHPO_4 \cdot 2H_2O$) due to the presence of a large amount silanol and molecular water content [61, 62], whereas the resonance peak around 0 ppm could be attributed to pure calcium orthophosphate phase [63]. The presence of Q_p^1 species is most apparent from the 70S10C spectra, which confirms the possible P–O–Si bonding existence in this glass composition. Recently, it is [64] suggested that this signal is derived from P–O–Si moieties stemming from P clusters at spots of silica-based pore-wall surface.

In the 63S20C glass sample, the deconvoluted ^{31}P spectra reveal that the major resonance band occurs at 0 ppm and that it is attributed to calcium orthophosphate (Q_p^0) species. The two weak bands corresponding to the chemical shifts at −1.7 ppm and 1.6 ppm again signify the presence of both monetite and brushite phases, respectively. This may be rationalized by suggesting that the introduction of Ca is associated with concurrently enhanced water adsorption and hence form hydrated calcium phosphate structure.

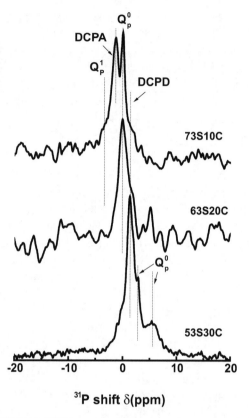

Figure 3.5 ^{31}P MAS-NMR spectra of representative pristine sol–gel glasses and glass-ceramic (DCPA: dicalcium phosphate anhydrous; DCPD: dicalcium phosphate dihydrate).

It is interesting to note from the ^{31}P spectra of the 53S30C glass-ceramic sample that we observe the five distinct chemical shift values between −1 ppm and 5.5 ppm. The obtained chemical shift values for individual band positions and the percentage of areas under the peak are listed in Table 3.2. The chemical shifts around −0.9 ppm and 1.3 ppm corresponding to monetite and brushite phases, i.e., the dehydrated or anhydrous dicalcium hydrogen phosphate amount is more in comparison to brushite phase due to the scarce amount of molecular water and silanol groups in the glass-ceramic sample. The chemical shift values around 0 ppm, 3 ppm, and 5.5 ppm are assigned to Q_p^0 species corresponding to orthophosphate but with different neighbors. More commonly, the chemical shift at 0 ppm is associated to calcium orthophosphate [65], whereas the chemical shift at 3 ppm is again same as that of crystalline calcium HAp [64], and in the present case, it is due to the presence of Ca–P clusters or HAp environment [66]. It is worth to mention that this chemical shift value is similar to that reported in Ca-rich bioactive glasses, where the appearance of the ^{31}P signal around 2.5–3.0 ppm is attributed to orthophosphate environment [58]. The observed chemical shift around 5.5 ppm, which lies between sodium orthophosphate (16 ppm) and calcium orthophosphate (0–3 ppm), suggests that the phosphate exists as mixed sodium–calcium orthophosphate environment, i.e., both types of modifier ions are associated with the orthophosphate units. Thus, the ^{31}P MAS-NMR spectra analysis underlines the fact that phosphorus is conventionally thought to enter as a network former; however, here we find that it exists as isolated orthophosphate structural units.

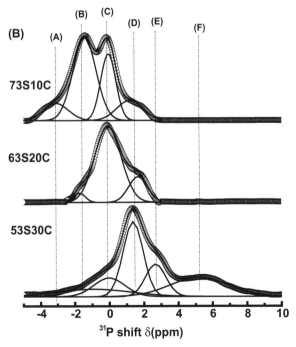

Figure 3.6 ^{31}P spectra for pristine glasses and glass-ceramic after deconvolution into various Q species for phosphate. (Open circle represents the obtained spectra, and solid line represents deconvoluted spectra).

Table 3.2 Relative populations (expressed as percentages) of Q_p^n species obtained after deconvolution of the obtained spectra

Sample	Q_p^1 (A) δ (ppm)	Area (%)	Monetite (B) δ (ppm)	Area (%)	Q_p^0 (C) δ (ppm)	Area (%)	Brushite (D) δ (ppm)	Area (%)	Q_p^0 (HAp) (E) δ (ppm)	Area (%)	Q_p^0 (Na, Ca) (F) δ (ppm)	Area (%)
73S10C	−3.1	11	−1.4	50	0	25	1.2	14				
63S20C			−1.6	19	0	71	1.6	10				
53S30C			−0.9	13	0	38	1.4	9	2.7	15	5.2	25

The bioactivity of glass and glass-ceramics has been carried out by analyzing both the as-prepared glass powder and the supernatant SBF solution. The *in vitro* apatite formation and changes in the chemical composition, i.e., the ion concentration of the SBF solution after soaking the samples for different durations, were systematically monitored. The chemical composition of the SBF solution after soaking was monitored by ICP-AES analysis as a function of soaking time. The chemical concentrations of individual ionic species released in the SBF solution for 73S10C and 53S30C samples after soaking in SBF for 12 h, 1 day, and 3 days at the ppm level are calculated and reported in Fig. 3.7. It is evident that the Ca species initially (12 h) show rapid release followed by a decrease in concentration with time, and this behavior becomes more prominent in the 73S10C glass sample than in the 53S30C glass-ceramic sample. Thus, the initial increase in Ca^{2+} concentration could be due to Ca^{2+} release from the soaked sample. The subsequent decrease in the Ca^{2+} concentration with the immersion time is attributed to the formation of the amorphous calcium phosphate layer over hydrated silica-rich layer, forming negative potential at the glass surface, necessary to incorporate Ca^{2+} ions from the SBF solution to the surface and form crystalline Ca–P. The change in P concentration follows a similar trend in both samples, but the 53S30C sample releases less phosphorus in comparison to the other two glass samples, which show that apatite formation ability becomes difficult in this composition. The reason for the slow ion dissolution of Ca and P in the 53S30C sample is the presence of different phases and phase boundaries. Therefore, the dissolution of ionic species depends solely on the type of phase in which ions are trapped as well as their textural nature. All the compositions show an increase in Si concentration during the soaking period, which indicates that Si continuously leached from the MBGs through the breakdown of the mesoporous structure and further leads to the degradation of bioactive glass.

The immersion of samples in the SBF solution also results in an appreciable increase in Na^+ concentration, which indicates that there is an exchange of Na^+ ions with the H_3O^+ present in the fluid. Therefore, the mesoporous glass samples accelerate the formation of Si–OH groups on their surface within 12 h of the dissolution than the glass-ceramic sample, where the rate of Na^+ ion dissolution is significantly slow, which again indicates the rate of reaction kinetics of apatite formation. The amount of consumption of calcium and phosphate in the solution was calculated by subtracting the value after 3 days of dissolution from the 12 h dissolution and compared in Fig. 3.8. This consumption data indicate the amount of CaP deposited on the 73S10C glass surface after 3 days, i.e., the Ca–P ratio of around 1.65 is very close to the stoichiometric Ca–P ratio of HAp (1.67). On the other hand, the 53S30C glass-ceramic sample shows a strong deviation (the estimated Ca–P ratio of 1.45) from the stoichiometric Ca–P ratio, which suggests the formation of the more meta-stable CaP phase other than HAp compared to the 73S10C glass sample.

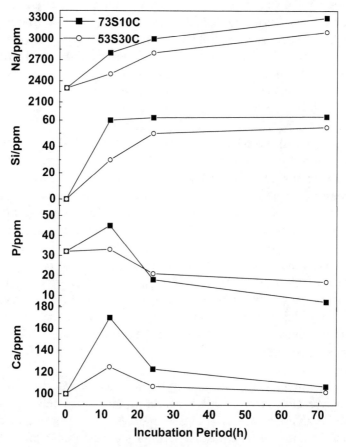

Figure 3.7 Ca, P, Si, and Na concentrations in SBF as a function of soaking time for 73S10C and 53S30C.

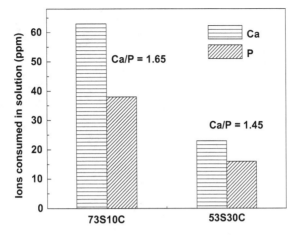

Figure 3.8 Total consumption of Ca and P for 73S10C and 53S30C.

Additionally, we have carried out transmission electron microscopy (TEM) and selected area electron diffraction (SAED) analysis of the *in vitro* bone mineralization process on glass and glass-ceramic sample surface. Micrographs depicted in Fig. 3.9 show crystalline aggregates composed of needle-like crystal formation on the surface of glass and glass-ceramic samples. The SAED pattern analysis reveals the (002), (211), and (222) reflections corresponding to apatite phase having with d-spacing of 0.34 nm, 0.28 nm, and 0.19 nm, respectively [67]. The observed phases are more visible in the case of 73S10C glass than other glass and glass-ceramic samples. More specifically, the glass-ceramic sample shows the less or absence of reflections corresponding to the apatite phase. The diffuse maxima in the observed XRD pattern of the glass-ceramic sample are manifested by the presence of nanocrystallites with random orientation.

Based on the above analysis, we can now provide more qualitatively the possible origin of the formation of ACP followed by crystalline HCA layer in the studied MBGs and highlight the various factors that influence the apatite formation ability. The glass compositions chosen in the present work are close to the known bulk 45S5 Bioglass; however, the local structural units present in the glass and glass-ceramic compositions are very much different as revealed by FTIR and NMR techniques. Therefore, we suggest that the formation of ACP has significant influence on the local

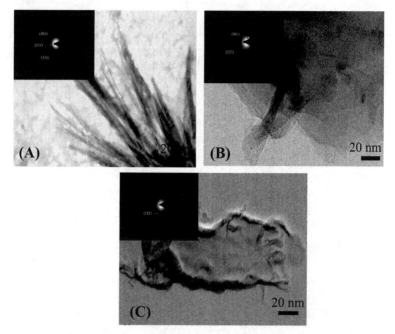

Figure 3.9 TEM micrographs and corresponding SAED patterns of (A) 73S10C, (B) 63S20C, and (C) 53S30C after soaking in SBF for 3 days.

structure of the given glass and glass-ceramics in addition to their textural properties. The structure of bulk 45S5 Bioglass has been characterized by the highly depolymerized fragment with various kinds of anionic species in which silicate species are found to be Q^3, Q^2, and Q^0 and phosphate species are found to be in orthophosphate form. The present structural investigations reveal that the silicate structure is less depolymerized (large Q^4 units) with a minor amount of orthophosphate species due to two reasons: (i) the modifier content is slightly less than 45S5 and (ii) intrinsically the sol–gel prepared samples have relatively less depolymerized nature than the corresponding bulk counterpart. Thus, the bioactivity of melt-derived glasses is mainly determined by their chemical composition, whereas the apatite formation in the MBGs is mainly controlled by their textural properties. Secondly, the MBGs contain a high amount of silanol density, which could play a crucial role on ACP formation than the composition of the glass. There are few reports to understand the effect of textural properties on bioactivity in different types of glasses and, in particular, much work has been carried out in the SiO_2–CaO–P_2O_5 glass system. Thus, we suggest that the obtained high Ca–P ratio in the present work stems from the superior textural property as well as the presence of different anionic species in the glass network structure.

In bulk glass samples, changing the chemical composition brings out the change in the local atomic structure of the glass alone without any modification in the textural properties. The 45S5 Bioglass comprises a significantly larger amount of network modifier than that present in MBGs, thereby possessing a depolymerized network built up of primarily Q^2 silicate units (and to a lesser extent Q^3 units). Hill [68] proposed that the optimal network connectivity (N_c) for MDBGs with $N_c \leq 2$ (45S5 Bioglass has an N_c of 1.9), but the same condition, was not good for sol–gel glasses, which are bioactive even with N_c around 3.7. In the present work, we find that the 73S10C glass is bioactive and shows a high Ca–P ratio close to the 45S5 glass composition. Since MBGs are characterized by high surface area and mesoporosity, it leads to the complex nature of the local atomic structure with a wide range of Q species. Although the NMR studies on these glasses show the different types of Q^n species with high N_c (around 3.7), still they exhibit enhanced bioactivity due to the presence of a large amount of hydroxyl groups.

Therefore, it is observed from the present investigation that there exists a large variation in the local atomic structure of MBG and MDBG. The ^{29}Si NMR evidenced the presence of Q^4 and $Q^3(OH)$ species after immersion in SBF although these species were not present in the as-prepared bulk glass samples [69]. The emergence of $Q^3(OH)$ and Q^4 species shows the replymerization of silanols to form a silica-rich layer and evidences the dissolution of glassy network. On the other hand, in the case of MBGs, the surface is enriched with silanols and physisorbed water molecules. Therefore, it has been proposed that the mechanism involved in HCA formation in MBGs will be similar to the Hench mechanism but with some important differences. Hence, the porosity and silanol density not only accelerate the first three stages of surface reaction but may also circumvent them. In addition to these crucial factors controlling the mechanism of bioactivity in MBGs, the presence of CaP cluster is also considerable.

Remarkably, we find that the substitution of CaO for SiO_2 inhibits the apatite formation on the surface of mesoporous glass. Clearly, it is distinguishable that the bioactivities of glass samples are higher than that of glass-ceramics, which consist of two phases—apatite-like and wollastonite. This type of apatite/wollastonite (A/W) has been widely studied in several aspects and is said to be bioactive [70]. However, an earlier study indicated that the apatite growth ability of glasses is higher than that of A/W glass-ceramics [71] due to the structure of wollastonite, in which Ca^{2+} is readily linked to the silicate network with less solubility [72]. Therefore, the presence of these two phases decreases the ion-dissolution process and hence leads to the slow rate of apatite formation. Secondly, the textural property of glass-ceramics is less superior to that of the glasses under study, which implies less amount of silanol formation in the glass-ceramic, which are nucleating sights for apatite and further slowing down of ion leaching process. Ultimately, all these factors result in the poor degree of apatite crystallization in glass-ceramics.

References

1. Lewandrowski, K. U., Gresser, J. O., Wise, D. L., and Trantolo, D. J. (2000) Bioresorbable bone graft substitutes of different osteoconductivities: A histologic evaluation of osteointegration of poly(propylene glycol-co-fumaric acid)-based cement implants in rats. *Biomaterials*, **21**, pp. 757–764.
2. Giannoudis, P. V., Dinopoulos, H., and Tsiridis, E. (2005) Bone substitutes: An update. *J. Care Injury*, **36**, pp. S20–27.
3. Salzman, N. P., Psallidopoulos, M., Prewett, A. B., and O'Leary, R. (1993) Detection of HIV in bone allografts prepared from AIDS autopsy tissue. *Clin. Orthop. Relat. Res.*, **292**, pp. 384–390.
4. Carter, G. (1999) Harvesting and implanting allograft bone. *J. Aorn.*, **70**, pp. 660–670.
5. Jones, J. R., Lin, S., Yue, S., Lee, P. D., Hanna, J. V., Smith, M. E., and Newport, R. J. (2010) Bioactive glass scaffolds for bone regeneration and their hierarchical characterisation. *Proc. Inst. Mech. Eng. H. J. Eng. Med.*, **224**, pp. 1373–1387.
6. Biomimetics, Tissue Engineering and Biomaterials, in National Institute of Dental Research Workshop, September 24–26, 1996.

7. National Institutes of Health. *Improving Medical Implants Performance through Retrieval Information: Challenges and Opportunities*, NIH Technology Assessment Conference, Jan 10–12, 2000.
8. Hutmacher, D. W., Schantz, J. T., Lam, C. X. F., Tan, K. C., and Lim, T. C. (2007) State of the art and future directions of scaffold-based bone engineering from a biomaterials perspective. *J. Tissue Eng. Regen. Med.*, **1**, pp. 245–260.
9. Habraken, W. J., Wolke, J. G., Jansen, J. A. (2007) Ceramic composites as matrices and scaffolds for drug delivery in tissue engineering. *Adv. Drug Deliv. Rev.*, **59**, pp. 234–248.
10. Zhou, H., and Lee, J. (2011) Nanoscale hydroxyapatite particles for bone tissue engineering. *Acta Biomater.*, **7**, pp. 2769–2781.
11. Hench, L. L., Splinter, R. J., Allen, W. C., and Greenlee, T. K. (1971) Mechanisms of interfacial bonding between ceramics and bone. *J. Biomed. Mater. Res. Symp.*, **2**, pp. 117–143.
12. Pantano, C. G., Clark, A. E., and Hench, L. L. (1974) Multilayer corrosion films on Bioglass® surfaces. *J. Am. Ceram. Soc.*, **57**, pp. 412–413.
13. Wu, C., Chang, J., and Xiao, Y. (2011) Mesoporous bioactive glasses as drug delivery and bone tissue regeneration platforms. *Ther. Deliv.*, **2**, pp. 1189–1198.
14. Abou Neel, E. A., Chrzanowski, W., and Knowles, J. C. (2008) Effect of increasing titanium dioxide content on bulk and surface properties of phosphate-based glasses. *Acta Bioamter.*, **4**, pp. 523–534.
15. Hench, L. L. (1991) Bioceramics: From concept to clinic. *J. Am. Ceram. Soc.*, **74**, pp. 1487–1510.
16. Acros, D., Lopez-Noriega, A., Ruiz-hernandez, E., Terasaki, O., and Vallet-Regi, M. (2009) Ordered mesoporous microspheres for bone grafting and drug delivery. *Chem. Mater.*, **21**, pp. 1000–1009.
17. Vallet-Regi, M. (2006) Ordered mesoporous materials in the context of drug delivery systems and bone tissue engineering. *Chem. Eur. J.*, **12**, pp. 5934–5943.
18. Yan, X., Yu, C., Zhou, X., Tang, J., and Zhao, D. (2004) Highly ordered mesoporous bioactive glasses with superior *in vitro* bone-forming bioactivities. *Angew. Chem. Int. Ed.*, **43**, pp. 5980–5984.
19. Garcia, A., Cicuendez, M., Izquierdo-Barba, I., Acros, D., and Vallet-Regi, M. (2009) Essential role of calcium phosphate heterogeneities in 2D-hexagonal and 3D-cubic SiO_2–CaO–P_2O_5 mesoporous bioactive glasses. *Chem. Mater.*, **21**, pp. 5474–5484.

20. Salinas, A. J., Merino, J. M., Gil, J., Babonneau, F., and Vallet-Regi, M. (2007) Microstructure and macroscopic properties of CaO–SiO$_2$ PDMS hybrids for use in implants. *J. Biomed. Mater. Res.*, **81B**, pp. 274–282.
21. Zhang, K., Yan, H. W., Bell, D. C., Stein, A., and Francis, L. F. (2003) Effects of materials parameters on mineralization and degradation of sol–gel bioactive glasses with 3D-ordered macroporous structures. *J. Biomed. Mater. Res. B*, **66A**, pp. 860–869.
22. Jones, J. R., and Hench, L. L. (2003) Regeneration of trabecular bone using porous ceramics. *Curr. Opin. Solid State Sci.*, **7**, pp. 301–307.
23. Baeza, A., Izquierdo-Barba, I., and Vallet-Regi, M. (2010) Biotinylation of silicon-doped hydroxyapatite: A new approach to protein fixation for bone tissue regeneration. *Acta Biomater.*, **6**, pp. 743–749.
24. Izquierdo-Barba, I., Ruiz-Gonzalez, L., Doadrio, J. C., Gonzalez-Calbet, J. M., and Vallet-Regi, M. (2005) Tissue regeneration: A new property of mesoporous materials. *Solid State Sci.*, **7**, pp. 983–989.
25. Yan, X., Huang, X., Yu, C., Deng, H., Wang, Y., and Zhang, Z. (2006) The *in vitro* bioactivity of mesoporous bioactive glasses. *Biomaterials*, **27**, pp. 3396–3403.
26. Lopez-Noriega, A., Acros, D., Izquierdo-Barba, I., Sakamoto, Y., Terasaki, O., and Vallet-Regi, M. (2006) Ordered mesoporous bioactive glasses for bone tissue regeneration. *Chem. Mater.*, **18**, pp. 3137–3144.
27. Salonen, J., Laitinen, L., Kaukonen, A. M., Tuura, J., Björkqvist, M., Heikkilä, T., Vähä-Heikkilä, K., Hirvonen, J., and Lehto, V. P. (2005) Mesoporous silicon microparticles for oral drug delivery: Loading and release of five model drugs. *J. Control. Release,* **108**, pp. 362–374.
28. Acros, D., and Vallet-Regi, M. (2010) Sol–gel silica-based biomaterials and bone tissue regeneration. *Acta Biomater.*, **6**, pp. 2874–2888.
29. Parez-Pariente, J., Balas, F., and Vallet-Regi, M. (2000) Surface and chemical study of SiO$_2$P$_2$O$_5$CaO(MgO) bioactive glasses. *Chem. Mater.*, **12**, pp. 750–755.
30. Murugavel, S., Vaid, C., Bhadram, V. S., and Narayana, C. (2010) Ion transport mechanism in glasses: Non-arrhenius conductivity and non-universal features. *J. Phys. Chem.*, **B114**, pp. 13381–13385.
31. Vaid, C., Murugavel, S., Khashyap, R., and Tandon, R. P. (2012) Synthesis and *in vitro* bioactivity of surfactant template mesoporus sodium silicate glasses. *Micro. Meso. Mater.*, **159**, pp.17–23.
32. Brunner, T. J., Grass, R. N., and Stark, W. J. (2006) Glass and bioglass nanopowders by flame synthesis. *Chem. Commun.*, **7**, pp. 1384–1386.

33. Mallick, K. K. (2009) Freeze casting of porous bioactive glass and bioceramics. *J. Am. Ceram. Soc.*, **92**, pp. S85–S94.
34. Yang, J. M., Lu, C. S., Hsu, Y. G., and Shih, C. H. (1997) Mechanical properties of acrylic bone cement containing PMMA–SiO$_2$ hybrid sol-gel material. *J. Biomed. Mater. Res. Appl. Biomater.*, **38**, pp.143–154.
35. Kamitakahara, M., Kawashita, M., Miyata, N., Kokubo, T., and Nakamura, T. (2003) Apatite formation on CaO-free polydimethylsiloxane (PDMS)-TiO$_2$ hybrids. *J. Mater. Sci. Mater. Med.*, **14**, pp. 1067–1072.
36. Sharp, K. G. (1998) Inorganic/organic hybrid materials. *Adv. Mater.*, **10**, pp. 1243–1248.
37. Novak, B. M. (1993) Hybrid nanocomposite materials—between inorganic glasses and organic polymers. *Adv. Mater.*, **5**, pp. 422–433.
38. Kokubo, T., and Takadama, H. (2006) How useful is SBF in predicting *in vivo* bone bioactivity? *Biomaterials*, **27**, pp. 2907–2915.
39. Huang, W., Day, D. E., Kittiratanapiboon, K., and Rahaman, M. N. (2006) Kinetics and mechanisms of the conversion of silicate (45S5), borate, and borosilicate glasses to hydroxyapatite in dilute phosphate solutions. *J. Mater. Sci. Mater. Med.*, **17**, pp. 583–596.
40. Huang, W., Rahaman, M. N., Day, D. E., and Li, Y. (2006) Mechanisms of converting silicate, borate, and borosilicate glasses to hydroxyapatite in dilute phosphate solutions. *Phys. Chem. Glasses Europ. J. Glass Sci. Technol. B*, **47**, pp. 647–658.
41. Yao, A., Wang, D. P., Huang, W., Rahaman, M. N., and Day, D. E. (2007) *In vitro* bioactive characteristics of borate-based glasses with controllable degradation behavior. *J. Am. Ceram. Soc.*, **90**, pp. 303–306.
42. Fu, Q., Rahaman, M. N., Fu, H., and Liu, X. (2010) *In vivo* outcomes of tissue-engineered osteochondral grafts. *J. Biomed. Mater. Res.*, **95A**, pp. 164–174.
43. Brown, R. F., Day, D. E., Day, T. E., Jung, S., Rahaman, M. N., and Fu, Q. (2008) Growth and differentiation of osteoblastic cells on 13-93 bioactive glass fibers and scaffolds. *Acta Biomater.*, **4**, 387–396.
44. Fu, Q., Rahaman, M. N., Bal, B. S., Brown, R. F., and Day, D. E. (2008) Mechanical and *in vitro* performance of 13-93 bioactive glass scaffolds prepared by a polymer foam replication technique. *Acta Biomater.*, **4**, pp. 1854–1864.
45. Wheeler, D. L., Stokes, K. E., Park, H. M., and Hollinger, J. O. (1997) Evaluation of particulate bioglass in a rabbit radius ostectomy model. *J. Biomed. Mater. Res.*, **35**, pp. 249–254.

46. Wheeler, D. L., Stokes, K. E., Hoellrich, R. G., Chamberlain, D. L. S., and McLoughlin, W. (1998) Effect of bioactive glass particle size on osseous regeneration of cancellous defects. *J. Biomed. Mater. Res.*, **41**, pp. 527–533.

47. Oonishi, H., Hench, L. L., Wilson, J., Sugihara, F., Tsuji, E., Kushitani, S., and Iwaki, H. (1999) Comparative bone growth behavior in granules of bioceramic materials of various sizes. *J. Biomed. Mater. Res.*, **44**, pp. 31–43.

48. Fu, Q., Saiz, E., Rahaman, M. N., and Tomsia, A. P. (2011) Bioactive glass scaffolds for bone tissue engineering: State of the art and future perspectives. *Mat. Sci. Eng.*, **C31**, pp. 1245–1256.

49. Fu, H., Fu, Q., Zhou, N., Huang, W., Rahaman, M. N., Wang D., and Liu, X. (2009) *In vitro* evaluation of borate-based bioactive glass scaffolds prepared by a polymer foam replication method. *Mater. Sci. Eng. C*, **29**, pp. 2275–2281.

50. Midha, S., Kim, T. B., van den Bergh, W., Lee, P. D., Jones, J. R., and Mitchell, C. A. (2013) Preconditioned 70S30C bioactive glass foams promote osteogenesis *in vivo*, *Acta Biomater.*, **9**, pp. 9169–9182.

51. Lebecq, I., Desanglois, F., Leriche, A., and Follet-Houttemane, C. (2007) Compositional dependence on the *in vitro* bioactivity of invert or conventional bioglasses in the Si–Ca–Na–P system. *J. Biomed. Mater. Res., Part A*, **83A**, pp. 156–168.

52. Saravanapavan, P., Jones, J. R., Pyrce, R. S., and Hench, L. L. (2003) Bioactivity of gel-glass powders in the CaO–SiO$_2$ system: A comparison with ternary (CaO–P$_2$O$_5$–SiO$_2$) and quaternary glasses (SiO$_2$–CaO–P$_2$O$_5$–Na$_2$O). *J. Biomed. Mater. Res., Part A*, **66A**, pp. 110–119.

53. Aina, V., Bonino, F., Morterra, C., Miola, M., Bianchi, C. L., Malavasi, G., Marchetti, M., and Bolis, V. (2011) Influence of the chemical composition on nature and activity of the surface layer of Zn-substituted sol-gel (bioactive) glasses *J. Phys. Chem. C*, **115**, pp. 2196–2210.

54. Izquierdo-Barba, I., Acros, D., Sakamoto, Y., Terasaki, O., Lopez-Noriega, A., and Vallet-Regi, M. (2008) High-performance mesoporous bioceramics mimicking bone mineralization. *Chem. Mater.*, **20**, pp. 3191–3198.

55. Pedone, A. J. (2009) Properties calculations of silica-based glasses by atomistic simulations techniques: A review. *J. Phys. Chem. C*, **113**, pp. 20773–20784.

56. Tilocca, A., and Cormack, A. N. (2007) Structural effects of phosphorus inclusion in bioactive silicate glasses. *J. Phys. Chem. B*, **111**, pp. 14256–14264.

57. Vaid, C., Murugavel, S., Das, C., and Asokan, S. (2014) Mesoporous bioactive glass and glass-ceramics: Influence of the local structure on *in vitro* bioactivity. *Micro. Meso. Mater.,* **186**, pp. 46–56.
58. Elgayar, I., Aliev, A. E., Boccaccini, A. R., and Hill, R. G. (2005) Structural analysis of bioactive glasses. *J. Non-Cryst. Solids,* **351**, pp. 173–183.
59. Christiansen, S. C., Zhao, D. Y., Janicke, M. T., Landry, C. C., Stucky, G. D., and Chmelka, B. F. (2001) Molecularly ordered inorganic frameworks in layered silicate surfactant mesophase. *J. Am. Chem. Soc.,* **123**, pp. 4519–4529.
60. Mercier, C., Follet-Houttemane, C., Pardini, A., and Revel, B. (2011) Influence of P_2O_5 content on the structure of SiO_2–Na_2O–CaO–P_2O_5 bioglasses by 29Si and 31P MAS-NMR. *J. Non-Cryst. Solids,* **357**, pp. 3901–3909.
61. Tseng, Y.-H., Zhan, J., Lin, K. S. K., Mou, C.-Y., and Chan, J. C. C. (2004) High resolution ^{31}P NMR study of octacalcium phosphate. *Solid State Nucl. Magn. Reson.,* **26**, pp. 99–104.
62. Circuendez, M., Portoles, M. T., Izquierdo-Barba, I., and Vallet-Regi, M. (2012) New nanocomposite system with nanocrystalline apatite embedded into mesoporous bioactive glass. *Chem. Mater.,* **24**, 1100–1106.
63. O'Donnell, M., Watts, S. J., Law, R. V., and Hill, R. G. (2008) Effect of P_2O_5 content in two series of soda lime phosphosilicate glasses on structure and properties–Part I: NMR. *J. Non-Cryst. Solids,* **354**, pp. 3554–3560.
64. Leonova, E., Izquierdo-Barba, I., Acros, D., Lopez-Noriega, A., Hedin, N., Vallet-Regi, M., and Eden, M. (2008) Multinuclear solid-state NMR studies of ordered mesoporous bioactive glasses. *J. Phys. Chem.,* C**112**, pp. 5552 –6662.
65. Aguiar, H., Serra, J., and Gonzalez, P. (2011) Nanostructural transitions in bioactive sol–gel silicate glasses. *Int. J. Appl. Ceram. Technol.,* **8**, pp. 511–522.
66. Hayakawa, S., Tsuru, K., Ohtsuki, C., and Osaka, A. (1999) Mechanism of apatite formation on a sodium silicate glass in a simulated body fluid. *J. Am. Ceram. Soc.,* **82**, pp. 2155–2160.
67. Izquierdo-Barba, I., Colilla, M., and Vallet-Regi, M. (2008) Nanostructured mesoporous silicas for bone tissue regeneration. *J. Nanomater.,* **3**, pp. 1–8.
68. Hill, R. J. (1996) An alternative explanation of the bioactivity of bioglasses. *Mater. Sci. Lett.,* **15**, pp. 1122–1125.

69. Dietrich, E., Oudadesse, H., Floch, M. L., Bureau, B., and Gloriant, T. (2009) *In vitro* chemical reactivity of doped bioactive glasses: An original approach by solid-state NMR spectroscopy. *Adv. Eng. Mater.*, **11**, pp. B98–105.

70. Lin, K., Zhang, M., Zhai, W., Qu, H., and Chang, J. (2011) Fabrication and characterization of hydroxyapatite/wollastonite composite bioceramics with controllable properties for hard tissue repair. *J. Am. Ceram. Soc.*, **94**, pp. 99–105.

71. Ragel, C. V., Vallet-Regi, M., and Rodriguez-Lorenzo, L. M. (2002) Preparation and *in vitro* bioactivity of hydroxyapatite/sol–gel glass biphasic material. *Biomaterials*, **23**, pp. 1865–1873.

72. Coleman, N. J., Bellantone, M., Nicholson, J. W., and Mendham, A. P. (2007) Textural and structural properties of bioactive glasses in the system CaO–SiO$_2$. *Ceramics-Silikaty*, **51**, pp. 1–8.

Chapter 4

Biomaterials Based on Natural and Synthetic Polymer Fibers

C. S. Krishna Murthy and Biman B. Mandal
Biomaterial and Tissue Engineering Laboratory,
Department of Biosciences and Bioengineering,
Indian Institute of Technology Guwahati (IITG),
Guwahati-781 039, Assam, India
biman.mandal@iitg.ernet.in, mandal.biman@gmail.com

In the world of bioengineering tissues, the uses of fibers as biomaterials have come a long way due to their various advantages, including their robustness, degradation, and the inherent ability to mimic extracellular matrix architecture in three dimensions. On the basis of their origin, fibers can be broadly divided into two groups: natural polymer fibers and synthetic polymer fibers. Natural materials include collagen, silk, cellulose, and fibrinogen, while biodegradable or resorbable synthetic materials include degradable polyesters such as polylactide and polyglycolide as biomaterials.

The safety aspects of fiber related to its biological characteristics such as biodegradability, biocompatibility, and antigenicity are of immense importance for its use in biomedical applications. Noteworthy biomedical applications of natural fibers include

---

Trends in Biomaterials
Edited by G. P. Kothiyal and A. Srinivasan
Copyright © 2016 Pan Stanford Publishing Pte. Ltd.
ISBN 978-981-4613-98-9 (Hardcover), 978-981-4613-99-6 (eBook)
www.panstanford.com

reconstruction of drug delivery systems, dressings for burns/wounds, delivery of bioactive molecules, sustained drug delivery, and as base substrates or scaffolds for tissue engineering and cell culture systems. In tissue engineering, natural fibers have been widely used for reconstruction of bone substitutes, repair of tendons and ligaments, replacement of skin, and artificial engineering of blood vessels and valves. Further, superabsorbent fibers have been developed from acrylic copolymers that absorb up to 50 times their own weight in water. They are used in dressings for wounds that release a large volume of exudates. Resorbable fibers (polylactic acid and polyglycolic acid and their copolymers) are specially developed to retain their mechanical properties *in vivo* for a specified period. Others include copolymers of polycaprolactone and polydioxanone. These materials are specifically designed to function for a predetermined period of weeks to months before they degrade.

In this chapter, we look into various categories of these natural and synthetic fibers for widespread application as biomaterials. Their specific fabrication methods, processing, testing, evaluation, advantages, and disadvantages are thoroughly described.

4.1 Introduction

Synthetic and natural biomaterials are made of "polymer" chains. A polymer ("poly" means "many" and "mer" means "part") is a high molecular weight macromolecule that consists of chain-like repeating structural units of many smaller molecules called monomers. Several different monomers or combinations of the same monomers ("mono" means "one" and "mer" means "part") are joined by covalent bonds to form polymers. Many polymers are organic and contain carbon as their major constituent element. Combinations of two or more polymers have been proved beneficial as biomaterials in tissue engineering. Biopolymers include synthetic polymers and natural polymers. Natural polymers can be found in living plants and animals—for example cellulose from plants, silk protein from silk worms and spiders, polysaccharides, and polynucleotides—whereas synthetic polymers are synthesized in the lab through a series of chemical reactions. Examples of such polymers are polyglycolide, polylactide, etc.

Biomaterials play a very important role in the modern strategies of biomedical engineering, tissue engineering, and regenerative medicine as a supportive design for biophysical and biochemical environments that direct cellular function and behavior [6–8]. Such biomaterials provide three-dimensional support to the cells with biomolecular interaction to perform their functions, and they control the complex multicellular processes of tissue formation and regeneration [9].

Various natural polymers are used as biomaterials for biomedical applications [10–13]. Natural polymers are mostly electrospun or meltspun to produce stabilized structures of nano-/microfibers for biological experiments. Natural polymers mixed with a variety of solvents can produce nano-/microfibers of varying diameters. For example, chitosan can be electrospun into nanofibers by using acetic acid [14] as well as trifluoroacetic acid as solvents [10, 15]; silk fibroin is electrospun into nanofibers by using water [12, 16], formic acid [17–19]; collagen nanofibers are produced by using 1,1,3,3,3-hexaflouro-2-propanol (HFP) [13, 20–22]; hyaluronic acid (HA) nanofibers are produced by using hydrochloric acid solution [23, 24]. Similarly, chitin nanofibers are produced with HFP solvents [18, 25].

In the early 20th century, researchers showed little interest in the development of polymer synthesis from glycolic acid, lactic acid, and other α-hydroxy acids because of their higher instability that make them undesirable for long-term applications. However, in the last four decades, the same unstable nature of these polymeric materials leading to biodegradation has been found to be very important for biomedical applications, and it began with the first approved medical use of biodegradable sutures in the 1960s [26, 27]. Since then, several biomedical devices based on glycolic acid, lactic acid, poly(ε-caprolactone) copolymers and homopolymers, poly(trimethylene carbonate) copolymers, and polydioxanone have been widely used for biomedical applications [28], including drug delivery [29, 30].

For any biomedical device, the physician basically opts for an implant that does not require a second surgery for its removal; biodegradability of these materials thus offers an additional advantage over stainless steel implants, which may have inclination toward re-fracture upon implant removal. For example, during the biological process of bone healing, the load is taken by the stainless steel and not by the original healing bone, leading to stress shielding. But the use of degradable polymer will slowly change the possession of the load to the healing bone [27, 31].

The criteria for selecting any polymer for biomaterial application should, among many, depend on its degradation time and mechanical properties so that it provides sufficient mechanical strength and does not collapse until the damaged tissue heals completely. An ideal polymer for biomaterial application should be nontoxic and easily sterilizable, show minimal inflammatory response, have agreeable shelf life, and metabolize inside the body after fulfilling its purpose without leaving any traces of the material [27].

The versatile properties of synthetic polymers have inspired several researchers to produce fiber materials suitable for biomedical and biotechnological applications. Fibers that are in the range of tens of microns in diameter appear to respond similar to a two-dimensional substrate with asymmetrical development of adhesion receptors leading to cell attachment and proliferation. These matrices have tremendous applications in tissue engineering, such as the reconstruction of engineered constructs like urinary bladder [32] and as scaffolds for neural stem cell regeneration [33]. Technologies such as electrospinning, melt spinning, and wet spinning allow polymer processing (for both natural and synthetic) and fiber formation down to 10 nm scale [34]. Some difficulties with various spinning techniques of nano-/microfibers include placing cells within the nanofibrillar structure with pore spaces smaller than the normal cell diameter. By some additional techniques, the network should be produced *in situ*, surrounding the cells, without damaging them. Additionally, rational design principles are utilized to control fiber morphology and scaffold architecture [35].

This chapter focuses on the aspects of different natural and synthetic polymeric fibers as biomaterials and the basic mechanisms underlying the techniques/methods of processing polymer fibers. The chapter also provides insight into the application of these polymer fibers for the fabrication of medical implants and devices. Several aspects related to polymeric fibers such as characterization methods/techniques, mechanical testing, evaluation, and various biomedical applications such as implant devices and artificial organs have also been discussed.

4.2 Polymer Fiber Materials

Polymer materials for manufacturing fibers can be both natural and synthetic. Some of these polymers that have been successfully used to fabricate fibers are discussed in the following subsections [36, 37].

4.2.1 Natural Polymer Fiber Materials

Even before the advent of plastic and synthetic polymers, nature provided several polymers that play vital roles to support life. For example, DNA and RNA present in the genes are involved in life processes. Natural polymers such as polypeptides (protein polymers like silk and fibrinogen) and polysaccharides (sugar polymers like chitin and chitosan) are made from carbon and hydrogen. Clinically, collagen and fibrin are widely used, and the Food and Drug Administration (FDA) has approved these polymer-based matrices as wound-dressing material to treat burns and chronic wounds and also as tissue sealants. Naturally derived materials represent more pertinent and valuable models for biomedical and material engineers to derive engineering principles and create artificial materials with similar biological function, making them the best choice for biomedical and biotechnological applications (Table 4.1) [38, 39].

Electrospinning, melt spinning, and wet spinning are majorly used to synthesize fibers from natural polymer materials. Natural materials tend to be more desirable to be fabricated into fibers than synthetic materials as the former typically have better interactions with cells and the total biological system due to their bioactive properties. These properties have been shown to promote wound healing in part due to their enhanced biocompatibility [40]. Such materials include collagen, silk fibroin, and polysaccharide-based materials such as chitosan, HA, and alginate. Compared to synthetic materials, natural materials' multifunctional properties allow them to have enhanced performance in biological systems [41]. Natural materials have the inherent capacity for binding cells since they carry specific protein cues such as RGD (glycine/arginine/aspartic acid) sequences [42]. Recently, electrospinning of natural polymers mainly from chitosan, silk fibroin, collagen, gelatin, and elastin has been reported [11–13, 43–46].

Table 4.1 List of natural polymer materials and their uses in various biological applications

Natural polymer	Application	Reference
Cellulose, cellulose acetate	Affinity membrane, adsorptive membranes/felts	[47, 48]
Chitosan/poly(ethylene oxide)	Wound healing, tissue-engineered scaffold, drug delivery	[49]
Collagen/chitosan	Biomaterials	[50]
Fibrinogen	Wound healing	[51]
Gelatin, gelatin/polyaniline	Scaffold for wound healing and tissue engineering	[44, 52]
Hyaluronic acid	Medical implant	[23]
Silk, silk/poly(ethylene oxide), silk fibroin	Biomedical applications, nanofibrous scaffolds for tissue engineering and wound healing	[53–56]

4.2.1.1 Silk

Silk is a natural biopolymer produced by insects such as silk worms (mulberry: *Bombyx mori*; non-mulberry: *Antheraea assama, A. mylitta, Philosomia ricini*, etc.) and spiders (*Araneus ventricosus, Nephila clavipes*, etc.). Silk from silk worms primarily consists of hydrophobic fibers like fibroin and hydrophilic glue like sericin proteins (Fig. 4.1). Depending on the source from where it is derived, its molecular weight is typically between 20 and 400 kDa (fibroin: 20–395 kDa; sericin: 24–400 kDa) [57, 58]. Apart from being biodegradable, biocompatible, and nontoxic, silk fibroin protein exhibits minimal inflammatory and immunogenic responses and offers good mechanical strength and thermal stability [59–66]. The primary structure of fibroin consists of a repeated sequence of amino acids, (Gly-Ser-Gly-Ala-Gly-Ala)$_n$, which are arranged in antiparallel β-sheets. The higher glycine and alanine content contributes to the tight packing of the β-sheets and provides tensile strength and rigid structure to the silk fibroin [61, 67–70]. The silk structure is controlled by chemical treatment, annealing, and compression to improve biodegradability and mechanical strength. [71]. Silk fibroin protein is blended with different synthetic or natural polymers to achieve higher mechanical strength and degradability [72]. When silk fiber is dissolved in a solvent, it gets denatured by the stretching of the hydrogen bonds between the fibers. Dissolved silk protein could be processed into different morphologies by using various techniques. During the final curing, insoluble β-sheet structures of the silk fiber are formed [73]. The rich crystalline β-sheet structure of silk gives high toughness and stiffness to fabricate biomaterials for load-bearing applications [60]. The amphiphilic nature of silk allows post-processing into various biomedical applications, including the fabrication of three-dimensional scaffolds [63]. Silk has also got FDA approval for its use in some medical devices [74].

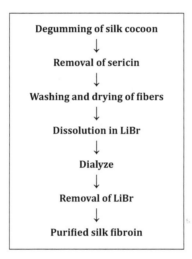

Figure 4.1   Flow chart depicting the conventional method of extraction of silk fibroin protein from cocoons of *Bombyx mori* silk worm.

4.2.1.2 Chitin and chitosan

Chitosan is a natural polysaccharide derived from chitin. It is found naturally in fungi (*Mucor rouxii, Aspergillus nidulans*), yeast, insects (cuticle, ovipositors, beetle cocoon), squid (*Ommastrephes* sp, loligo stomach wall), centric diatoms (*Thalassiosira fluviatilis*), algae, and in the exoskeleton of sea crustaceans (crab shell, shrimp shell) (Fig. 4.27). Depending on the source, the molecular weight of chitin is typically between 300 and 1000 kDa. Chitin is the second most abundant polysaccharide in the world, after cellulose. Chitosan is naturally found in limited quantities in some fungi. Chitosan is biodegradable, biocompatible, nontoxic, hemostatic, and a natural antibacterial agent [75]. Chitosan promotes wound healing, stimulates the immune system, and prevents infection within the surgical site [40, 76]. It has the ability to bind to fats, proteins, cholesterols, metal ions, and tumor cells. This allows chitosan to be used as a chelating agent in various biomedical and industrial applications [77, 78]. Chitin has a homogeneous chemical structure made up of 1-4 linkages of 2-acetamido-2-deoxy-β-D-glucopyranose. Chitosan is derived from chitin through chemical or enzymatic treatments. The deacetylation of chitin to at least 50% of the free amine form, which has a heterogeneous chemical structure made up of both 1-4 linked 2-acetamido-2-deoxy-β-D-glucopyranose as well as 2-amino-2-deoxy-β-D-glucopyranose, generally produces chitosan with 70% to 95% deacetylation [75].

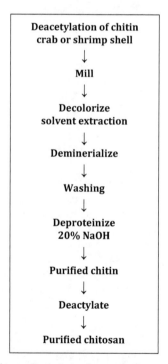

Figure 4.2 Flow chart depicting the transformation of chitin from crustacean cuticle to chitosan [75].

The properties of chitosan are greatly affected by the conditions under which it is processed. Chitosan is a highly reactive polysaccharide. The degree of deacetylation controls the amount of positively charged free amino groups and along with the hydroxyl group gives chitosan its functionality. The important aspect of chitosan's processability is that its positive charge allows it to have many electrostatic interactions with negatively charged molecules and for side group attachment. It affects the crystallinity and directly relates chitosan's ability to solubilize in acidic aqueous solutions [41, 79].

Degradation is an important property to be considered during processing, so that end applications can be designed accordingly [79]. By controlling the amount of deacetylation during processing, the degradation rate of chitosan could be engineered. Its degradation rate is inversely proportional to the degree of crystallinity and consequently the amount of deacetylation. Thermal degradation at above 280°C rapidly breaks down the polymer chains. Degradation can also be controlled by enzymatic methods. Hydrolytic enzymes like lysozyme naturally degrade chitosan by releasing amino sugars, which can be easily processed and released through the metabolic system of the body [76].

4.2.1.3 Collagen

Collagen is a natural biodegradable polymer. It is the major structural protein present in animals, and it exists in sheets and fibrillar forms consisting of triple helix of three protein chains. Type I fibrillar collagen makes up to 25% of the protein mass of the human body. It is abundant and easily isolated from human tissues, and it is most often used in biomedical devices. Collagen fibrils have high tensile strength, and they are hemostatic, which makes them useful for various applications. Collagen, a fibrous protein, is the most abundant protein in the extracellular matrix (ECM) preserved across species [80] and, therefore, elicits the weakest immune response. For thousands of years, natural collagen sutures were implanted effectively. The sutures composed of animal sinew are almost entirely collagen. Bovine collagen is one of the most abundantly available xenogenic materials widely used in biomedical applications [81]. Collagen is commonly obtained from the lower layer of bovine hide from the bovine corium. Xenogenic bovine collagen is nonimmunogenic. Sometimes, even though it is so well preserved, immune response may occur in humans who are hypersensitive to it [82–84] and allergic to beef [85]. Generally, cleaning with detergents and terminal sterilization by ethylene oxide gas or gamma irradiation reduce the minimal immune response levels, lower than the synthetic meshes [86–89]. Immunogenic reactions from xenogenic collagen are caused by differences in telomeres, or repetitive end "buffer" sections of the collagen molecules. The sections can be removed without harming the structural integrity of the protein molecule. However, the processing breaks apart the collagen fiber, greatly reducing the strength of any biomaterial made from it.

Collagen extracted from animal tissues can be both water soluble and water insoluble. Water-soluble collagen quickly resorbs water and degrades rapidly in the physiological moist environment, and hence the preparation of materials and devices is very limited. However, water-insoluble collagen is commonly used in the manufacturing of medical devices. It is grounded, purified, and harvested in the powdered form, which is later processed to manufacture materials and devices. Collagen polymer cannot be melt processed, so it is processed by evaporating water from collagen suspensions. Collagen fibers and tubes are formed by extruding collagen suspensions into buffered aqueous solutions at pH 7.5 [85]. Collagen devices are often chemically modified or crosslinked to make them less hydrophilic and to reduce degradation. Viswanadham and Kramer have showed the water content of untreated collagen hollow fibers with 15–20 μm thickness and 400 μm outer diameter as a function of humidity. The water plasticizes the collagen, lowering both the yield strength and modulus. Methods using ultraviolet (UV) radiation for crosslinking the fibers increase the modulus of the fibers [90].

4.2.1.4 Hyaluronic acid

Hyaluronic acid is a biocompatible, biodegradable, bioresorbable, and bioactive natural polymer. It is a linear polysaccharide and belongs to the glycosaminoglycans family. It mainly consists of repeating units of glucuronic acid and N-acetylglucosamine and is specifically present in the ECM of conjunctive tissues such as dermis and epidermis [91]. HA fibers are produced by the wet spinning and electrospinning methods by coagulating the aqueous solution of sodium hyaluronate in various solvents such as concentrated acetic acid [23, 43]. HA has good viscoelasticity with greater power of water retention, which makes it useful in skin hydration [92, 93]. It is actively used in tissue engineering, wound healing, implant materials, arthritis treatment, tissue scaffolds, and drug delivery applications [46, 95].

4.2.1.5 Gelatin

Gelatin is a biodegradable, biocompatible natural polymer. It is of two types: gelatin type A and gelatin type B. Both are extracted and processed from collagens by acidic pretreatment and alkaline pretreatment, respectively [11, 94, 96]. It is commercially available at cheaper cost and is widely used in medical and pharmaceutical applications, as wound dressings [97, 98], carriers for drug delivery [99–101], and sealants for vascular prostheses [102–104]. However, gelatin nano-/microfibers produced by electrospinning, wet spinning, and melt spinning are not common [20, 105]. Furthermore, it has difficulty in dissolving as colloidal sol at or above 37°C in water and coagulates into gels at room temperature. Special treatment such as crosslinking of gelatin is required for tissue engineering purposes. Gelatin is mixed with synthetic polymers to make new bioartificial polymeric materials [106], which may reduce the problem of cytotoxicity due to the use of chemical crosslinking reagent and provide good biocompatibility as well as improved mechanical and chemical/physical properties. Electrospinning of gelatin and poly(ε-caprolactone)(PCL) produces gelatin/PCL composite fibrous scaffolds whose properties are evaluated and compared with its constituent materials [11, 46].

4.2.1.6 Fibrinogen

Fibrinogen is a natural polymer and a soluble fibrillar protein secreted into the plasma by hepatocytes in high amounts [107, 108]. Fibrinogen plays an important role in coagulation cascade and as a bridging molecule for platelet aggregation [109]. Fibrinogen and its derivative fibrin also serve as a tentative ECM in some injured tissues [110] and are involved in wound healing [111], tissue regeneration [112], inflammatory cell response [113], and angiogenesis [114]. Fibrin as well as fibrinogen is used as tissue sealant, scaffold for cellular therapies [115], and in tissue engineering of skin, vascular grafts, cartilage, and bone [116–118]. Bowlin and coworkers first successfully prepared electrospun native fibrinogen nanofibers [51]. It is also shown that cells such as fibroblasts can easily colonize the electrospun fibrinogen scaffolds [119]. The fiber stability in the aqueous environments, cellular recognition, and locomotor activity toward the spatial orientation of fibrinogen nanofibers significantly influence the cytoskeleton organization in the confluent cell layers and the secretion of extracellular fibronectin matrix by the cells [108].

4.2.2 Synthetic Biodegradable Polymers

Synthetic materials are often used to fabricate fibers because of their low cost, greater availability, and a wide range of acceptable solvents. Synthetic polymers offer more advantages over natural materials. They can be tailored to give a wider range of desirable properties with more predictable uniformity over natural materials. In addition, a reliable source of raw materials is obtained with synthetic polymers with no concerns for immunogenicity [27, 28]. Synthetic fibers have enhanced mechanical properties such as tensile strength and modulus. They are easily engineered for specific applications and have simple purification methods as compared to natural materials [41].

Factors that influence the mechanical performance of biodegradable synthetic polymers depend on the selection of the monomer, its initiator, processing conditions, and the availability of additives. These will consecutively affect the polymer's molecular weight and its distribution, end groups, hydrophilicity, crystallinity, glass transition and melt temperatures, sequence distribution (random versus blocky), the presence of residual additives or monomers, and biodegradation [27]. Polymers are synthesized to obtain hydrolytically unstable linkages in the backbone to accomplish biodegradation. The common preferable chemical functional groups are anhydrides, amides, esters, and orthoesters.

4.2.2.1 Absorbable/resorbable polymer fiber materials

Absorbable materials can be described as biomaterials purposefully designed to be degraded *in vivo* into nonharmful byproducts and are finally excreted (e.g., CO_2 and water from lactic acid) or metabolized (e.g., tricalcium phosphate in polylactide composite). Earlier, bioresorbable polymers were widely used in sutures (Dexon, Vicryl, Maxon, Monocryl, PDS, Polysorb, Biosyn) [120–124]. Nowadays their uses have expanded for various biomedical applications, for example implants for trauma surgery plates, membranes [125], drug carriers [126], delivery devices in gene therapy [127], three-dimensional porous scaffolds, tissue or organ substitutes, and tissue-engineered implants [128–131].

Various functional and commercial aspects are considered while selecting resorbable polymers as implants. They should be biocompatible, have good mechanical properties in the physiological environment lasting till fracture healing, and biodegradable within the adequate time (their daily release of minimal byproducts into the surrounding tissue and through body's excretory routes). Commercially, they should be produced in high-purity reproducible batches, without affecting their final properties, and should have a legally certified clinical history to be used in patients, for example sutures.

Polyhydroxy acids occupy the main position among all resorbable polymers for implants. They include polyglycolide and poly(L-lactide) copolymers based on glycolide, L-lactide, DL-lactide, L/DL-lactide, ε-caprolactone, and trimethylene carbonate. In addition, tyrosine-carbonate-based experimental polymers have been used as implant materials.

A brief overview is given on biodegradable synthetic polymers currently used in orthopedic fixation devices (rods, pins, tacks, screws, ligaments) and wound closure (staples, sutures). Most of the commercially available synthetic biodegradable devices are polyesters composed of copolymers or homopolymers of lactide and glycolide. There are also other products made from copolymers of ε-caprolactone, trimethylene carbonate, and polydioxanone [27].

4.2.2.2 Polyglycolide

Polyglycolide is the simplest linear aliphatic polyester. Davis and Geck developed DEXON (marketing name), the first completely synthetic absorbable suture by using polyglycolic acid (PGA) [132, 133]. The dimerization of glycolic acid will lead to the synthesis of glycolide monomer. The polymerization of glycolide by ring opening produces high molecular weight materials with the presence of 1–3% residual monomer (Fig. 4.3). PGA has a higher degree of crystallinity (45–55%) and is not soluble in most of the organic solvents with the exceptions of highly fluorinated organic solvents such as hexafluoroisopropanol. PGA has a glass transition temperature of 35–40°C and a high melting point of 220–225°C [133]. Fibers made from PGA show high strength and modulus. PGA fibers are too stiff for sutures except as braided materials. PGA sutures lose almost 50% and 100% of their strength after 2 weeks and 4 weeks, respectively, and are completely absorbed within 4–6 months [133]. Copolymerization of glycolide with other monomers reduces the stiffness of the resulting fibers [27].

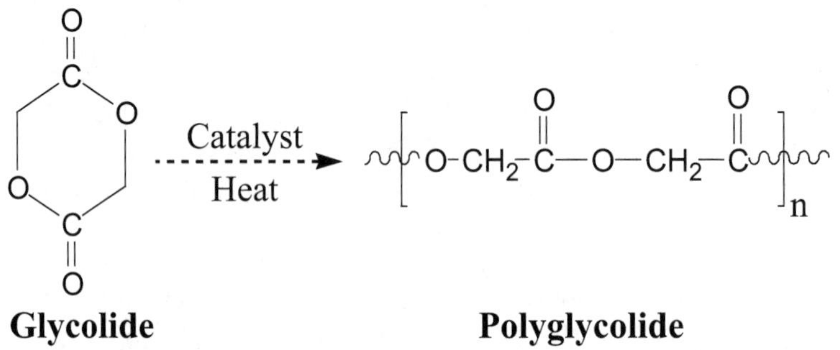

Figure 4.3 Synthesis of polyglycolide [27].

4.2.2.3 Polylactide

Polylactide is the cyclic dimer of lactic acid. It exists in D and L optical isomers. L-lactide isomer is the naturally occurring form. D-lactide and L-lactide are synthetically blended to DL-lactide. The polymerization methods of both glycolide and lactide are similar (Fig. 4.4). The L-lactide homopolymer (LPLA) exists in semicrystalline form. PGA and LPLA both exhibit high tensile strength, high modulus, and low elongation, which make them more applicable for load-bearing orthopedic fixation and sutures. Poly(DL-lactide) (DLPLA) is an amorphous polymer. It exhibits greater elongation, lower tensile strength, and rapid degradation in less time, making it more attractive for drug delivery applications [27]. Poly(L-lactide) has 37% crystallinity with a glass transition temperature of 60–65°C and melting point of 175–178°C [134, 135]. The degradation time of LPLA is much slower and requires more than 2 years to be completely absorbed [136]. L-lactide copolymers with glycolide or

DL-lactide were prepared to disorder the crystallinity of L-lactide, thereby enhancing the degradation process [26, 27, 133].

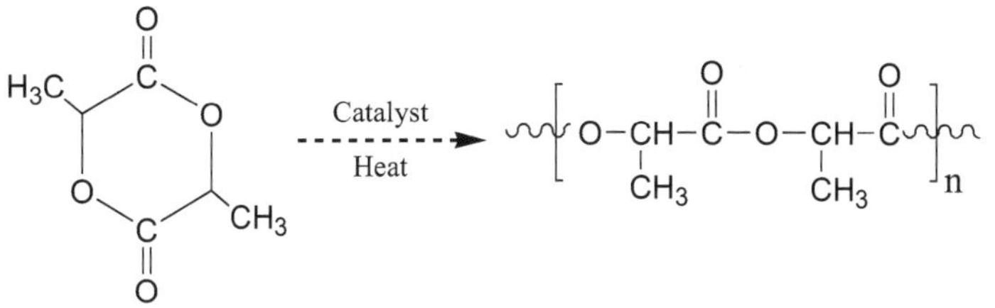

Lactide → **Polylactide** (Catalyst, Heat)

Figure 4.4 Synthesis of polylactide [27].

4.2.2.4 Poly(ε-caprolactone)

The ring opening (rearrangement) polymerization of ε-caprolactone, a cyclic ester, produces the semicrystalline polyester poly(ε-caprolactone) polymer with a glass transition temperature of −60°C and melting point of 59–64°C (Fig. 4.5) [27]. The homopolymer degrades in 2 years. The ε-caprolactone copolymers with DL-lactide produce materials with greater degradation rates [137]. A block copolymer of glycolide and ε-caprolactone is marketed in the name of MONOCRYL; the monofilament suture is traded by Ethicon [27, 132].

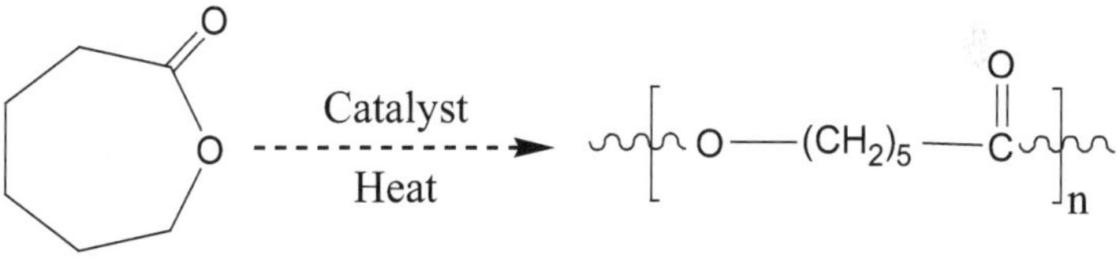

ε-carprolactone → **poly(ε-carprolactone)** (Catalyst, Heat)

Figure 4.5 Synthesis of poly(ε-caprolactone) [27].

4.2.2.5 Polydioxanone

The first clinically tested synthetic monofilament suture is synthesized by the ring opening polymerization of p-dioxanone and is being marketed in the name of PDS® by Ethicon (Fig. 4.6). Fraction fixation orthopedic devices such as absorbable pin composed of polydioxanone are also being marketed [138]. Polydioxanone has

55% crystalline property with a glass transition temperature of −10°C to 0°C. Polydioxanone did not show any toxic effects upon implantation [133].

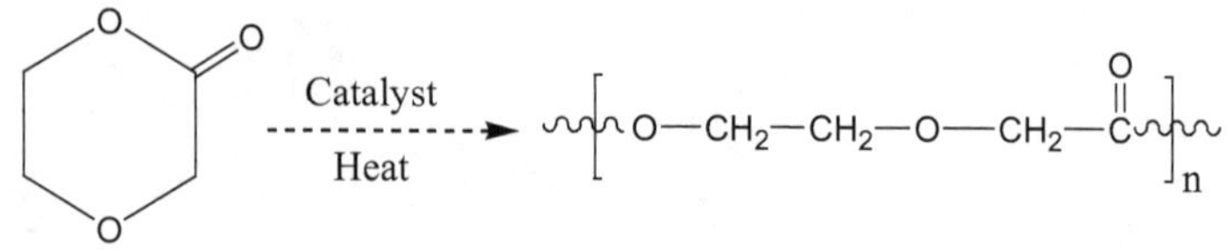

***p*-dioxanone** **polydioxanone**

Figure 4.6 Synthesis of polydioxanone (a polyether-ester) [27].

4.2.2.6 Poly(lactide-co-glycolide)

Polyglycolide and poly(L-lactide) (PLLA) are used as base materials, and the two monomers are copolymerized to improve the homopolymer properties (Fig. 4.7). Glycolide copolymers of L-lactide and DL-lactide are used for drug delivery applications and devices [27]. Copolymers of 25–70% glycolide and L-lactide are amorphous; the other monomer disrupts the regular polymer chain [26]. Several compositions of poly(DL-lactide-co-glycolide) copolymers are used for suture anchors, screws, and plates for surgical repairs [138]. Polyglyconate, a copolymer of glycolide and trimethylene carbonate (TMC), is used for manufacturing sutures, tacks, and screws [138]. Polyglyconate copolymers are prepared as A-B-A block copolymers in the ratio of 2 : 1 (glycolide : TMC) with pure glycolide blocks (A) at the end and having central glycolide-TMC block (B) in the middle. These materials have improved flexibility than pure PGA and can easily be absorbed within 7 months [133]. Nowadays, resorbable fixation devices made from copolymers and homopolymers of lactide, glycolide, caprolactone, trimethylene carbonate, and *p*-dioxanone have been commercialized (Table 4.2) [27, 138].

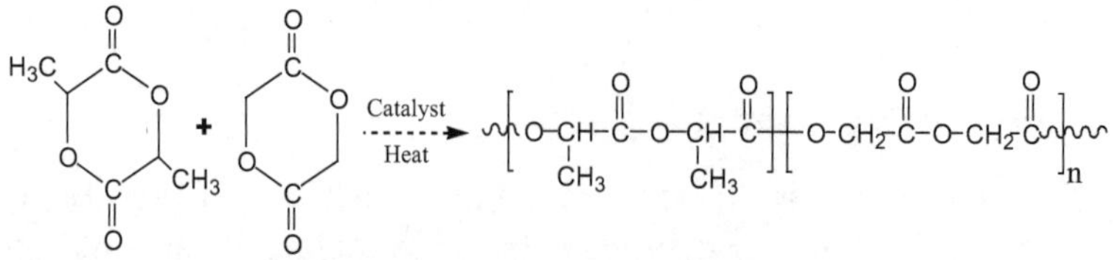

Lactide **Glycolide** **Poly(lactide-co-glycolide)**

Figure 4.7 Synthesis of poly(lactide-co-glycolide) [27].

Table 4.2 Physical properties of biodegradable polyesters

Polymer material	Tensile strength (psi)	Elastic modulus (ksi)	Specific gravity	T_m (°C)	T_g (°C)
Polyglycolide	10,000	1000	1.53	225–230	35–40
Poly(L-lactide)	8000–12,000	600	1.24	173–178	60–65
Poly(DL-lactide)	4000–6000	400	1.25	Amorphous	55–60
50/50 Poly(DL-lactide-co-glycolide)	6000–8000	400	1.34	Amorphous	45–50
Polycaprolactone	3000–5000	50	1.11	58–63	−65 to −60

Source: Adapted from Birmingham Polymers, Inc [133, 139].

4.3 Fabrication and Processing of Polymer Fibers

Biodegradable polymers are melted for processing and then synthesized into fibers, molded parts, and rods. The final parts are then extruded, molded by injection or compression, and solvent spun or casted. In some cases, the primary processing is subsequently followed by machining into the final parts. During processing, the molecular weight may be decreased due to the presence of moisture and the hydrolytic degradation of the polymeric bonds altering the final polymer properties. To avoid these problems, the polymer is dried by a vacuum dryer or resorption, circulating dry air to remove the moisture before thermal processing to reduce degradation. The influence of several processing factors on degradation is studied by Michaeli and von Oepen as well as Middleton *et al.* [140–142]. Care must be taken while drying the polymers above room temperature. For example, amorphous polymer pellets should be dried exclusively at room temperature because above their glass transition temperature, the polymer pellets may start to fuse.

Most of the resorbable synthetic polymers have been synthesized by ring opening polymerization. A thermodynamic equilibrium exists between the reverse reaction and the polymerization reaction, which results in the formation of the monomer. Higher processing temperature pushes the equilibrium to depolymerization and results in the formation of monomer during the extrusion or molding process. The excess monomer acts as a plasticizer, which may alter the mechanical properties and degradation kinetics by catalyzing the hydrolysis of the device [133]. Temperature, shear rate, residence time, and moisture content in the machine may strongly affect the degradation during processing [140, 141]. The higher shear rates and long residence time may result in increase in polymer degradation at even lower temperatures. For optimal processing, the mildest possible conditions without moisture are recommended. In many situations, this is difficult due to very high molecular weight polymers being extruded or molded into small fibers or parts in the devices. High temperature and pressure are required to reduce melt viscosity and to enable the flow of polymer through the small orifices to fill a mold or create the fiber. Repetitive molding or extrusion is needed to get the desired properties of the final part for various applications [27, 142]. Several processing operations for the fabrication of polymer fibers such as electrospinning, melt spinning, wet spinning, and filament molding are discussed in the following subsections.

4.3.1 Electrospinning

Electrospinning is a technique for producing fibers from polymer solutions or melts using electrostatic forces that result in the creation of fibers with diameters ranging from nanometers to micrometers. It has the ability to produce fibers that are far smaller than those produced by conventional means of fiber spinning, such as wet spinning, dry spinning, melt spinning, and gel spinning. This technology has been known since the early 1930s [36, 143]. Since then it has been employed for a wide range of organic polymers and has proven to be capable of being electrospun with nanosized diameters. Its wide range of applications include filtration of subatomic particles, composite reinforcement, multifunctional membranes, tissue engineering scaffolds, wound dressings, drug delivery, artificial organs, and vascular grafts [36, 37, 144–147]. Through electrospinning, biocompatible polymers can be spun into a nanosized mesh. This nanomesh has high surface area and a nanosized diameter that can be used to mimic a natural ECM. It also acts as scaffold allowing cells to adhere, proliferate, differentiate, and develop into functional tissue [41, 145, 146, 148]. Cells with an artificial ECM encourage tissue growth and therefore promote healing.

4.3.1.1 Electrospinning apparatus setup

The electrospinning apparatus mainly consists of a high voltage power supply, a needle or spinneret, and a grounded collector plate. Figure 4.8 shows the basic components needed to perform electrospinning. A syringe is filled with a polymer solution or melt, which is fed at a precisely controlled rate through a metallic needle or a spinneret into an electric field usually through a pump syringe or gravity. The polymer solution or melt travels from the syringe into a metallic needle, which is connected to a high voltage power supply between 1 and 30 kV. The conductive plate completes the circuit by creating a strong electric field between the needle and the plate. The grounded plate also collects the nanofiber web fabricated during the electrospinning process [37, 144].

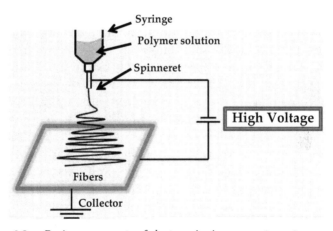

Figure 4.8 Basic components of electrospinning apparatus setup.

Fibers are produced due to static electric forces that manipulate a polymer solution while moving through an electric field generated by a high voltage power supply, between a conducting capillary (needle or spinneret) that holds a polymer solution and a grounded collector plate. The polymer solution is held in the shape of a droplet by surface tension, and as the electric field moves from the high voltage supply to the needle and then to the solution, a charge is induced on the surface of the droplet [37, 144]. An opposite charge (repulsion force), also formed at this time, pulls the droplet toward the electric field. As the electric field increases, the charge on the droplet also increases and causes the spherical droplet to deform into a conical shape. This is known as the Taylor cone. The force from the electric field influences the repulsion force until it overcomes the surface tension, and the charged solution is pulled into the electric field toward the grounded plate. As the solution, now a fluid filament, is pulled through the electric field, internal and external charges cause the liquid jet to be whipped around within the field as it accelerates toward the grounded collector. This action simultaneously causes the solvent to evaporate and the polymer chains within the solution to stretch and slide past one another. This whipping motion allows the fibers on the grounded collector to have diameters small enough to classify them as nanofibers [36, 37, 144–146, 149].

4.3.1.2 Electrospinning processing parameters

Numerous variable parameters are involved in producing a fiber mesh of uniform nanosized diameter. All variables, including solution, processing, and ambient parameters, must be considered during fiber formation. Solution parameters include the type, concentration, and molecular weight of the polymer as well as the type, concentration, viscosity, conductivity, and surface tension of the solvent. Processing parameters include electric field strength, distance between needle and collector, flow rate, and shape and movement of the collector. Ambient parameters include temperature and humidity [36, 149]. Solution viscosity is an important parameter and is dependent on polymer concentration. A minimum amount of polymer chain must be used for any spinning mechanism to allow chain entanglement and fiber production [145, 146]. Increase in polymer concentration increases the number of polymer chains present in the solution. It also increases the viscosity to allow flow through the needle and prevents the Taylor cone and fiber formation [145, 146, 149]. This is particularly critical in polyelectrolyte polymers due to the "polyelectrolyte effect." Ambient parameters such as humidity and temperature also affect the electrospinning process. There is an inverse relationship between temperature and viscosity. Mit-Uppatham *et al.* (2004) have observed the effect of temperature on the electrospinning of polyamide-6 fibers ranging from 25 to 60°C [150]. Increase in temperature produces fibers with decreased diameter due to the decrease in the viscosity of the polymer solution, while high humidity encourages the discharge of the electrospun fibers on the collector plate [92, 144].

4.3.2 Melt Spinning

Melt spinning is a process to manufacture polymeric fibers. It is a stress-induced crystallization process. The melted form of the bulk polymer is pumped through a spinneret (die) consisting of numerous small holes (one to thousands) ranging in size from 0.3 to 2.5 mm in diameter, depending on the fiber size required (Fig. 4.9). The smelted fibers are cooled down to a temperature well below the melting point [1, 4]. Crystallization starts between the glass transition temperature (T_g) and the equilibrium melting point (T_m) [3]. At this stage, the high crystalline orientation of the polymer chains (crystallite threads) is achieved by stretching/drawing. During cold drawing, the fiber should be kept above the glass transition temperature. The fibers stretching in the molten and solid states favor the orientation of the polymer chains through the fiber axis, and finally the fibers are collected at the take-up wheel (rotating roller). Further, annealing (i.e., heating below the softening point) is accompanied to stabilize the properties and the final fiber structure by stress relief. The fiber is held uptight during heating and subsequent cooling events to prevent relaxation [1, 151].

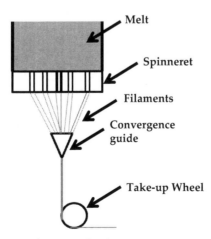

Figure 4.9 Process schematic of melt spinning operation.

Controlled stress is applied during or after crystallization to improve the strength and stretch resistance of polymers. X-ray diffraction (XRD) technique is used to study the molecular rearrangement caused by the "drawing" (stretching) operation. Examples of polymers that are melt spun in high volumes include polyethylene (PE) and poly(ethylene terephthalate) [4, 5].

4.3.3 Wet Spinning

Wet spinning, the oldest process, is used for polymers that are subjected to thermal degradation at higher temperatures. For example, when a polymer is encapsulated with a drug, a relatively low temperature wet spinning process is used. The polymers are dissolved in a solvent and extruded through the spinneret into a non-solvent in the chemical bath (the solvent is soluble in the chemical bath, while the polymer is not), causing the solution to coagulate into filament strands. This allows the fiber to precipitate and then solidify while emerging. The fibers are then taken out of the bath, washed to remove all solvents and non-solvents, and dried before being reeled on the spindles (Fig. 4.10) [152]. The fibers are spun in the chemical bath, and hence the process gets its name as wet spinning [2]. This method is used for spinning acrylic, aramid, and rayon fibers [5].

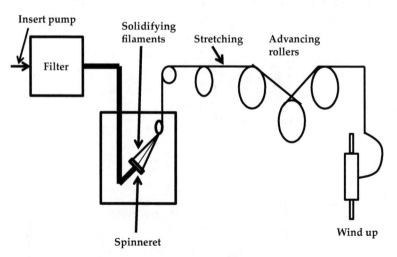

Figure 4.10 Process schematic of wet spinning operation [152].

4.3.4 Filament Winding

Filament winding is used to manufacture parts with fractions of high fiber volume and controlled fiber orientation. The fiber tows are immersed in a resin bath and impregnated with medium or low molecular weight reactants. The fiber is then directly wound around a rotating mandrel in a controlled mold to form the shape of the part. Subsequently, the resin is cured by using heat. After process, the mandrel (mold core) can be removed or left as an integral part of the instrument (Fig. 4.11) [4].

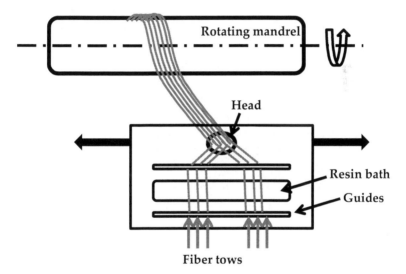

Figure 4.11   Process schematic of filament winding operation [4].

4.4 Characterization, Testing, and Evaluation of Fibers

4.4.1 Characterization of Polymer Fibers

Fabrication of polymeric fibers by electrospinning has attracted the attention of researchers due to their morphological characteristics. For the characterization of morphologic and geometric properties, techniques such as scanning electron microscopy (SEM), transmission electron microscopy (TEM), and atomic force microscopy

(AFM) are used [44, 153]. These nanofibers have high surface areas, small pore sizes, and these characteristics can be modified through process parameters to suit the applications and needs of the three-dimensional forms [36]. The porosity and pore size of nanodiameter membranes are important for applications in filtration, tissue engineering, and protective fabrics [154]. The molecular structure of fiber can be characterized by X-ray photoelectron spectroscopy (XPS), Fourier transform infrared (FTIR) spectroscopy, and nuclear magnetic resonance (NMR) techniques [155].

The mechanical properties of nanofiber matrices are crucial for biomedical applications. Mechanical characterization is performed by applying tensile test loads to specimens of the electrospun ultrafine non-woven fiber mats. Mechanical characterization of nanofibers and nanowires is done by using nanoindentation, resonance frequency measurements, microscale tension tests, and bending tests. Young's modulus, tensile strength, and the strain at break are determined by performing tensile tests with single polymeric fibers. For example, the nanotensile testing system is used for determining the mechanical properties of single ultrafine polymeric fibers of polycaprolactone (PCL) [156]. The tension tests follow the macroscale standards and count for the least number of assumptions to predict the material properties. This method for fiber testing is useful for diameters up to 1 μm and can be performed until fiber failure [157]. The elastic properties of electrospun membranes are determined by AFM, with the help of a cantilever and tip assembly that is used for scanning the surface. By reading the cantilever deflection caused due to the repulsion of the tip on the contacting atomic shells with very slight contact, the sample atomic resolution can be determined [158]. The advanced version of this technique is the AFM phase imaging, used for the detection of changes in hardness and composition [159]. The AFM-based nanoindentation method is used to determine the elastic modulus of one-dimensional electrospun nanofibers and nanowires. However, it has certain limitations such as silicon tip shape of the nanoindenter, relative tip fiber configuration, the effect of roughness, fiber surface curvature, the adhesion force between the indenter with silicon AFM tip, and the biological sample [46, 160]. A three-point nanoscale bending test with an AFM tip is used to determine the elastic modulus of nanofibers. Elastic modulus increases with decrease in the fiber diameter due to the shearing of the fibrils within the nanofibers to the total deflection of the nanofibers. The alteration in mechanical properties is due to the orientation of the nanofibers at the time of its formation under severe strain forces [161]. Moreover, the nanofiber properties are dependent on both the processing parameters and characterization methods. In the case of electrospun fibers, small pores are formed due to the evaporation of the solvent, and these pores influence the mechanical properties (such as tensile strength and Young's modulus) of the fibers [37]. However, the physical properties of nanofibers seem to be inferior to that of similar thickness film and resin counterparts [36]. The lower crystallinity could result due to the rapid evaporation of the solvent, followed by rapid cooling at the final stages of the electrospinning process [37].

4.4.2 Mechanical Testing and Evaluation

Mechanical evaluations are done to check the robustness of the fabricated electrospun fiber matrices. When implementing a mechanical testing program, reference information may be obtained from the results of failure analysis, prior regulatory submissions, FDA documents, American Society for Testing and Materials (ASTM) standards, and Association for the Advancement of Medical Instrumentation (AAMI)/International Organization for Standardization (ISO) standards. Failure tests determine the tensile/ultimate properties of the material. Industrial tests of polymers are performed to simulate the end use. The failure tests are done by using a "tensile testing machine." The test material is molded or cut into standard dimensions in the form of a dumbbell (dog bone) or a square-shaped specimen depending on the instrument specification. The specimen is firmly clamped in the grips of the tensile testing machine. A gradually increasing tensile force (in N or lb) is applied on the material at a slow but constant speed until rupture occurs. The amount of deformation (extension) is then measured by an extensometer. The applied tensile force (stress) and the extension (strain) are simultaneously recorded. Finally, a "stress–strain curve" is plotted to evaluate the tensile properties. Tensile testing is conducted under the same conditions for various materials that give direct comparison of the strength properties of different materials [1].

In this chapter, we confine our discussion to only the mechanical testing of both natural and synthetic polymer biomaterials. First, we present a detailed discussion on various mechanical characterization methods such as uniaxial tests, ball burst tests, and biaxial tests, and then we discuss the assessment of fiber alignment to optimize the mechanical properties of biomaterials for individual applications.

Mostly, natural polymer materials are hyper-elastic, and they differ from linear elastic materials. The stress and strain relationship is initially linear, but the elastic modulus increases on an exponential function (Fig. 4.12). The relationship is described by a strain energy density function [162].

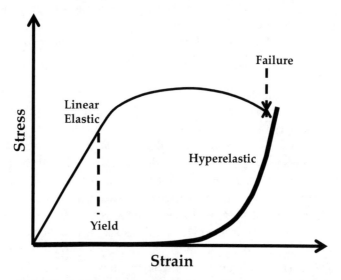

Figure 4.12 Typical hyper-elastic and linear elastic stress (σ)–strain (e) curves. When kept under the same strain, the linear elastic material experiences plastic yield and then failure, while the hyper-elastic material still deforms elastically [162].

4.4.2.1 Uniaxial tension and burst tests

The most conventional and simple mechanical tests for polymeric biomaterials are the uniaxial tensile and compression tests [162]. The specimen is stretched through cycles until rupture to examine the strength and tensile/Young's modulus. Generally, the specimen is given the shape of a dog bone or a dumbbell and then placed in a pair of separated grips (Fig. 4.13), which are controlled by the specimen strain. A few anisotropic materials are tested in several directions, and the obtained results are used to create the profile of a full material response. Uniaxial tensile and compression tests are performed for pseudo-elastic or elastic solid materials used in orthopedic applications (e.g., plastics and metals) whose material properties can be defined with only a few parameters. Soft compliant materials have the least compressive strength and are very difficult to grip and thus difficult to test. ASTM standard D3787-01 provides the specifications to test textile strength by implementing the "ball burst test," which is also adopted for testing soft tissues. Burst tests are useful for easy, quick, and accurate comparisons between various materials and treatment protocols [162].

Square or Dumbell shaped specimen

Figure 4.13 (a) Dumbbell- or dubbed dog-bone-shaped specimen mounted in the vice grips of a uniaxial testing platform. (b) Cutting of an anisotropic specimen in several directions [162].

For performing the burst test for biomaterials, a thin sheet of the sample is clamped into a ring, producing a taut sheet by carefully removing the slack without actually stretching the tissue. Then, at a slower rate, the spherical plunger is pushed through the biomaterial until the sphere ruptures it. Initially, with the minimal force, 10–20% of stretch can occur in the hyper-elastic tissues [162] (Fig. 4.14).

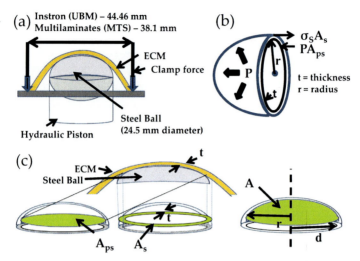

Figure 4.14 (a) Ball burst cage with its dimensions. (b) Free-body diagram of a thin-wall spherical pressure vessel. (c) Definition of areas used in analysis (ECM = extracellular matrix; A, A_s, and A_{ps} = different areas; t = initial thickness of the material; and r = effect radius). Reprinted from Ref. 199, Copyright 2005, with permission from Elsevier.

The maximum burst force and plunger travel distance at rupture can then be recorded, and these quantities can vary with the testing conditions. The ASTM standard provides the exact dimensions of the apparatus. The standard ball burst test employs a single cycle to rupture the biomaterial without preconditioning, because preconditioning conceivably replicates the state of a biomaterial when implanted and also greatly changes the material response [163] to loading. Upon implantation, the *in situ* biomaterials are subjected to the multiple mechanical stresses, including periodic, cyclic, and other varied loads occurring during everyday movement.

There are some limitations of the uniaxial tension test or the ball burst test. Uniaxial tests cannot detect the coupling that occurs within the tissue in different orientations to modify the reaction to strain or stress in a particular direction. In addition, the necking effect causes the tissue to undergo plastic deformation, which is similar to the Poisson's ratio in the elastic tissues. Burst tests combine the coupling effects into a single cycle test by radially stretching the material to failure, and hence the material never retains its original state [164]. These limitations are solved in the biaxial mechanical testing experiments [162].

4.4.2.2 Biaxial tests

Biaxial mechanical testing experiments are more complex but can assess and accurately replicate the stress states of the polymeric biomaterial inside the body. In biaxial tests, the two perpendicular axes are controlled independently, and preconditioning of the materials can be done to a biologically relevant degree. In 1948, Treloar *et al.* first developed biaxial tests by applying two independent strains in orthogonal directions by measuring both the strains independently [162]. These tests were performed on all the four edges of a square specimen while allowing the unconstrained changes of the specimen in the thickness direction. The major problem in these tests is to necessarily control each boundary to its degrees of freedom while leaving the other boundaries without restraint.

The biaxial test machine is similar to two uniaxial testing machines arranged perpendicular to each other (Fig. 4.15). Each axis is provided with independent actuators and a force gauge. Throughout the tests, a tissue is kept in a reservoir containing simulated body fluid maintained at physiological temperature. The square-shaped specimen is fastened to the actuators by using threads or sutures. The specimen is usually made in the square shape to get more accurate results [164]. The square shape provides the least distortion of the strain field, taken from the central (1/4) measurements [165] by safely ignoring the effects of stress concentrations. From the actuators bearing the load cells, the force and stress can be directly measured and calculated, respectively. Strain and deformation are subsequently calculated by setting a camera above the specimen and tracking four or nine dots on the surface of the specimen from the suture attachment points. An advanced computer-controlled machine can easily calculate the strain and precisely control the strain rate to get accurate results. Various modes of strain, cycling and loading are obtained by varying the rate of strain at both sides. The material can be tested at their known levels of physiological range of stresses and strains. The results of biaxial tests are useful for characterizing the mechanical properties and relative strengths and for determining the usability of a polymeric biomaterial for a specific application [164, 166].

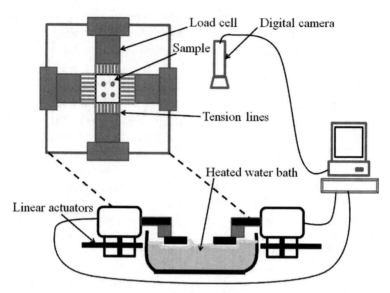

Figure 4.15 Diagrammatic representation of the biaxial testing system. A square-shaped specimen, represented with four dots for video strain measurement, is in a water bath immersion and several thin lines attached to four linear actuators [162].

4.4.2.3 Identifying fiber orientation

High interspecimen variability in biomaterials can often cause poor agreement between results of different sets of specimens. Natural polymer biomaterial specimens are composed of random fibers intricately connected into the framework of natural tissues. This variability can be reduced with the identification of the principal axis or the "strongest" axis. The principal axis is the direction in which the highest number of fibers are oriented, and the axis perpendicular to the principal axis is the weakest. Biaxial mechanical tests along these axes give a minimum and maximum strain–stress profile of the material. Generally, in natural materials, more number of fibers are aligned in the direction of the highest load [164]. An advanced biaxial testing machine allows the real-time feedback control of the strain rate while calculating both normal and shear strains [162, 167].

Synthetic materials are manufactured with exact specifications, and the determination of the principal axes and the degree of anisotropy is relatively simple. Microscopic inspection of tissues such as tendon and muscle can reveal the dominant fiber orientation. However, in many cases, the fibers are relatively small and challenging to discern. Fiber orientations in natural materials have been studied using several techniques such as electron microscopy [168] and standard polarized light microscopy [169]. These techniques are, however, time consuming and can acquire only local information.

Indirect methods can also be used to determine the principal material axes of natural biomaterials to identify the orientations of the fibers and the results can be verified against XRD [170]. Small-angle light scattering (SALS) [171], polarized light microscopy (PLM), and birefringence or double refraction methods can be used to qualitatively determine fiber directions [169, 172]. In the birefringence method, when a ray of light passes through a material, depending on the polarization of the light, it decomposes or splits into two rays having different velocities. When the rays are recombined in the microscope analyzer, constructive and destructive interference patterns are developed (Fig. 4.16). These interference patterns are measured to calculate the angle of the predominant internal structure of the material using sine of the phase difference:

$$\delta = 2\pi L(n_2 - n_1)/\lambda$$

where δ is the phase difference of the light, L is the thickness of the sample, n is the material refractive index, and λ is the wavelength of the light [173]. However, the rotating polarizer and additional algorithms are used to get more accurate measurements [173]. The assessment of the fiber orientations in a specimen is done before and/or after the mechanical testing to show the instant state of the material.

Recently, advanced PLM techniques have been used for the quantitative real-time assessment of fiber alignment in natural materials. This method enables the tracking and detection of the fiber orientations over large areas of the specimen by determining each pixel in the image [174]. The analyzer detects the sinusoidal oscillations by the light intensity passing through the birefringent material [162, 174]. Finally, with all the available data, the material modeling is done to predict the reaction of the material to different stresses applied.

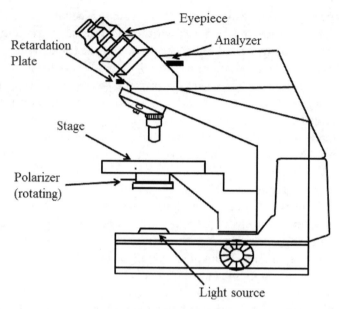

Figure 4.16 Diagrammatic representation of a polarized light microscope (PLM) showing the position of the specialized components. The birefringent material on the stage transmits the characteristic interference pattern of polarized light at each angle. For a complete picture of a biomaterial, the rotating polarizer facilitates the analysis of multiple angles.

Definitions:

Stress (σ) is the force applied per unit area of a specimen.

Strain (e) is the proportional (or percent) elongation.

Rupture strength is the stress at which complete breakage occurs.

Tensile strength (ultimate strength) is the maximum stress supported by the specimen during the test.

4.5 Applications of Polymer Fiber Material

Natural and synthetic polymers have been used in biomedical applications such as wound closings and wound dressings [73, 175, 178]. Polymeric fibers are used in a wide range of products, including sutures and implantable devices such as artificial extracellular matrices and vascular grafts [175]. Polymeric nanofibers have large surface area and could be engineered in various forms for applications such as tissue engineering scaffolds [76, 41, 146, 148], artificial organs [175], drug delivery [147, 178], multifunctional membranes [146], composite reinforcement [147], and filtration [75, 146].

4.5.1 Polymeric Fibers in Biomedical Applications

Natural and synthetic polymeric fibers are used for various biomedical applications. Smith *et al.* have produced the skin mask by wrapping electrospun fibers directly onto the skin surface to heal or protect the wounds [145]. Electrospun fiber mats used as drug delivery vehicles were made from either poly(ethylene-co-vinyl alcohol) (PEVA), PLA, or a 50 : 50 blend using a model drug such as tetracycline hydrochloride [176]. PLA is widely used for biomedical applications due to its good mechanical properties, biodegradability, and biocompatibility, and it can be easily processed by dissolving in common solvents [17]. Biomaterials can be used for constructing bone implants, cements, dental implants, cosmetic surgeries, heart valves, and vascular grafts [177]. Electrospinning can be used to create thin biocompatible fiber films along with silk-like polymer associated with cell-binding functionality (ECM proteins) for the useful coating design. The film can be deposited on the surface of prosthetic devices, which will facilitate the device for integration with the central nervous system and the body [145].

4.5.2 Extra Cellular Matrix Replacement

The nanofiber scaffolds mimic the structure and function of ECM in terms of morphology, chemical composition, and surface functional group. Techniques such as electrospinning, melt spinning, and wet spinning produce the polymeric material forms as artificial ECM [144, 148]. The artificial ECM sufficiently supports the mechanical and biological needs of the cells. Further, polymeric materials are used as interface materials or artificial replacements within the body and in various biomedical applications [21, 41, 49, 146]. Different tissues have different functions and are composed of variable sizes of cells, which will influence the cellular attachment, proliferation, and function of the cell. Therefore, for any type of cell, the pore size of the ECM should support the cell attachment and migration into the matrix while allowing nutrient flow and exchange of metabolic waste. After the formation and integration of the new tissue in the body, the artificial matrix should degrade safely within the biological system. The artificial matrix should be resistant to stress and strain [37, 41, 146].

Polymeric nanofibers can be aligned while collecting during the processing operation and can be formed into various three-dimensional structures, which could then drive migration of the cells [148]. Nanofibers can also be functionalized by surface and bulk modifications. Surface modifications using plasma treatment can influence the surface characteristics of the contacting cells [146], while bulk modification is done by incorporating bioactive molecules such as drugs, antibacterial agents, and growth factors into the polymer solution throughout the processed spun nanofibers [75, 147].

4.5.3 Drug Delivery

Biodegradable polymer solutions were spun into fibers that incorporate bioactive molecules for drug delivery devices [147, 178]. The rate of release of bioactive molecules depends on the degradation rate of the polymer, fiber thickness, and also on the ability of the body to absorb the drug [175]. The spinning operations help in creating the nanosized engineered material with controlled degradation rate and for the optimization of the encapsulated drug inside the body [75]. The electrospun fibers are incorporated with low molecular weight drugs such as cefazolin [178], ibuprofen [179], rifampin, paclitaxel [180], and itraconazole [181] and hydrophilic drugs such as tetracycline hydrochloride [176] and mefoxin [182].

Zhang et al. [183] have developed coaxially configured poly(ethylene glycol) (PEG) in poly(ε-caprolactone) (PCL) fibers encapsulating a model protein conjugate of fluorescein isothiocyanate and bovine serum albumin [183]. They have successfully demonstrated the smooth release of the drug over 5 days from the electrospun nanofiber mats. Coaxial electrospinning is also used to develop the drug-loaded polymer nanofibers for controlled and successful release of two different encapsulated drugs [46].

4.5.4 Wound Dressings

The pore size of the electrospun nanofiber mesh makes it an excellent candidate for wound healing, hemostatic devices, and burn treatment. Additionally, the mesh protects the damaged tissue from invading bacteria at the infected vulnerable wound site. The ideal wound dressing should allow transmission of water vapor, gaseous exchange, functional adhesion (i.e., selective adherence to the healthy tissue without adhering to the wound tissue) causing no pain to the patient, easily removable, and should be available at cheaper cost [46]. Electrospun nanofibers are similar to the natural ECM and also support the growth of new healthy tissue in the injured area, avoiding the formation of scar tissue and decreasing the healing time. The large surface area and high porosity help in the absorption of exudates and pus as well as fasten the wound healing process [75, 175]. Electrospun nanofibrous membrane can be used for wound dressing because of its higher porosity, which provides higher gas or oxygen permeability and protects the wound from dehydration and infections by acting as a good barrier. PCL nanofiber membranes and biocomposites support the growth of human dermal fibroblast and keratinocytes in skin tissue engineering [184]. PCL has resorption time of more than 1 year but is susceptible to enzymatic degradation [185], and nanofibrous membrane exhibits immediate, uniform adherence to the wet wound surface without the accumulation of any fluids and providing good support for wound healing [186].

4.5.5 Tissue Engineering Applications

Tissue engineering combines the principles of engineering and biological sciences for the development of artificial organs or the biological substitutes for the restoration, maintenance, and improvement of damaged or diseased tissue function. Biomaterials are used in tissue engineering for the fabrication of functional three-dimensional matrices for cell growth, proliferation, migration, and new tissue formation [7].

4.5.5.1 Scaffolds for tissue engineering

Biodegradable scaffolds are the temporary templates used in tissue engineering for cell seeding, proliferation, and differentiation, which are required for the regeneration of biological tissue or natural ECM. The three-dimensional scaffolds support cell–cell and cell–matrix [94]. The diameter of nanofibers produced by processing techniques such as electrospinning and melt spinning mimics the fibrils in ECMs for effective cell growth [187]. Natural polymers are biocompatible and have biofunctional motifs used for making nanofibrous scaffolds. Collagen, chitosan, silk protein, alginate, fibrinogen, hyaluronic acid, and starch are blended with synthetic polymers for the improvement of scaffold cytocompatibility [188, 189]. Several varieties of polymeric nanofibers have been used as scaffolds for dermal tissue engineering [186], arterial blood vessels [190], heart [191], nerves [192], cartilages, and bones [22]. Electrospun PLGA fiber mats have porosity greater than 90% with multiple focal adhesion points, and their high surface area makes them ideal scaffolds for cell attachment growth and proliferation for a wide variety of cell types. For example, the adherence and spreading of mouse fibroblasts are well studied on the orientation of PLGA nanofibers [14, 94]. The wet spinning technique improves porosity and cell migration [193]. The wet electrospinning method, a combination of electrospinning and wet spinning systems, could be used for the formation of three-dimensional spongiform nanofiber fabric with controlled fiber density. In this method, pure water is used as the wet spinning solvent whose surface tension significantly affects the fiber density; 50% and 99% tertiary-butyl alcohol (t-BuOH) and poly(glycolic acid) are used as polymer [46].

4.5.5.2 Scaffolds for vascular graft engineering

Synthetic polymers can be specifically modified to control degradation rate. They are biocompatible and offer high strength and elasticity. The FDA has approved the use of PGA, polylactide, and polydioxanone as the bioresorbable graft materials. Shum-Tim et al. [194] used polyglycolic acid and polyhydroxylalkanoate copolymer scaffolds embedded with a combination of ECs, SMCs, and fibroblast cells derived from lamb carotid arteries and implanted back into lamb aorta and showed 100% patency for 5 months [194]. Niklason et al. [195] constructed a graft using modified PGA scaffolds and bovine SMCs. The graft was subjected to pulsatile flow conditions, and after implantation into the porcine model, it showed 100% patency at 4 weeks [195]. The grafts showed burst pressures above

2000 mm Hg with no suggestive contractile response. Moreover, the breakdown products of biocompatible PGA fibers are acidic and cause immune response. Wu et al. [196] used porous polyglycolic acid–poly-L-lactic acid (PGA–PLLA) scaffolds and seeded them with human umbilical cord, EPC-derived EC phenotype cells, and human smooth muscle cells to form microvessels. The grafts are implanted *in vivo* in nude mice, and the formation of functional microvessels was observed after 7–10 days [196]. Gao et al. [197] used poly(glycerol sebacate) (PGS) films and scaffolds and seeded them with baboon endothelial progenitor cells (BaEPCs) and baboon smooth muscle cells (BaSMCs). BaSMCs were distributed throughout the construct *in vitro*, showed BaEPC adhesion and ECM synthesis, and maintained their cell-specific phenotypes [197]. Xu et al. [198] used PGA (polyglycolic acid) fiber mesh and SMCs. An elastic vessel wall was formed after 8 weeks under dynamic pulsatile stimulation of bioreactor and showed orientation of SMCs and collagenous fibers *in vitro* [198].

4.6 Conclusion

Biomaterials implanted inside the body have to interact well with the native tissue while performing their requisite functions without being physiologically rejected. To suit this purpose, several synthetic and natural polymeric materials were used to manufacture fiber-based medical devices. Natural sources such as silk, chitin, and chitosan were the first materials used for these procedures. In the early 20th century, synthetic materials occupied a huge market for fabricating medical implants. The exquisite mechanical properties of the implant materials should suit conveniently to the location in which they are implanted. For improvement in the applicability of these fibers, electrospinning and melt spinning techniques are widely used. Fibers were fabricated from the copolymerization of the polymeric mixtures for the attainment of enhanced physical and biological properties. Today, fiber-based biomaterials of all types (natural and synthetic) are being used for applications such as wound dressing as well as tendon and ligament repair. In addition to the fabrication and processing of polymer fibers, methods of characterization, examination of the morphological and mechanical properties of polymeric fibers, maintenance of the correct anisotropy, and the appropriate material response to mechanical stimuli should be studied. The empirical evidence obtained through simple burst or biaxial tests is needed for confirming the studies. Biomedical engineers use these data to determine the suitability of the implant and study mechanical failure when subjected to wear and tear inside the human body. Research should focus on the improvement of nanofiber properties and the scale-up of this process. Researchers have come a long way from the beginnings of natural and synthetic polymeric biomaterial implantation and have reached a point where the available technology converges with the knowledge of using modern tools to yield us valuable and accurate information. In the future, polymer biomaterials can prove to be invaluable substrates for designing medical implants for *in vivo* applications.

References

1. http://chemrat.com/ChemHog2/Polymer%20Chem_files/Polymer%20Notes05.DOC.
2. http://textilelearner.blogspot.com/2011/08/spinning-process-of-filament-yarns_8904.html#ixzz2YF5wtUmv
3. http://www.engr.utk.edu/mse/Textiles/Olefin%20fibers.htm
4. http://www.polymerprocessing.com/operations/filwind/index.html
5. http://www.uotechnology.edu.iq/teachers/UploadFolder/polymeric%20fibers%20Manufacturing.pdf
6. Peppas, N.A., and Langer, R. (1994) New challenges in biomaterials. *Science,* **263**, pp. 1715–1720.
7. Hubbell, J.A. (1995) Biomaterials in tissue engineering. *Nat. Biotechnol.,* **13**, pp. 565–576.
8. Langer, R., and Tirrell, D.A. (2004) Designing materials for biology and medicine. *Nature,* **428**, pp. 487–492.
9. Lutolf, M.P., and Hubbell, J.A. (2005) Synthetic biomaterials as instructive extracellular microenvironments for morphogenesis in tissue engineering. *Nat. Biotechnol.,* **23**, pp. 47–55.
10. Ohkawa, K., Cha, D., Kim, H., Nishida, A., and Yamamoto, H. (2004) Electrospinning of chitosan. *Macromol. Rapid Comm.,* **25**, pp. 1600–1605.

11. Zhang, Y., Ouyang, H., Lim, C.T., Ramakrishna, S., and Huang, Z.M. (2005) Electrospinning of gelatin fibers and gelatin/PCL composite fibrous scaffolds. *J. Biomed. Mater. Res. B: Appl. Biomater.,* **72**, pp. 156–165.

12. Li, L., and Hsieh, Y.L. (2006) Chitosan bicomponent nanofibers and nanoporous fibers. *Carbohyd. Res.,* **341**, pp. 374–381.

13. Zhong, S., Teo, W.E., Zhu, X., Beuerman, R.W., Ramakrishna, S., and Yung, L.Y.L. (2006) An aligned nanofibrous collagen scaffold by electrospinning and its effects on *in vitro* fibroblast culture. *J. Biomed. Mater. Res. Part A.,* **79A**, pp. 456–463.

14. Geng, X., Kwon, O.H., and Jang, J. (2005) Electrospinning of chitosan dissolved in concentrated acetic acid solution. *Biomaterials,* **26**, pp. 5427–5432.

15. Chong, E.J., Phan, T.T., Lim, I.J., Zhang, Y.Z., Bay, B.H., Ramakrishna, S., and Lim, C.T. (2007) Evaluation of electrospun PCL/gelatin nanofibrous scaffold for wound healing and layered dermal reconstitution. *Acta Biomaterialia,* **3**, pp. 321–330.

16. Jin, H.J., Chen, J.S., Karageorgiou, V., Altman, G.H., and Kaplan, D.L. (2004) Human bone marrow stromal cell responses on electrospun silk fibroin mats. *Biomaterials,* **25**, pp. 1039–1047.

17. Kim, S.H., Nam, Y.S., Lee, T.S., and Park, W.H. (2003) Silk fibroin nanofiber: Electrospinning, properties, and structure. *Polym. J.,* **35**, pp. 185–190.

18. Min, B.M., Lee, S.W., Lim, J.N., You, Y., Lee, T.S., Kang, P.H., and Park, W.H. (2004) Chitin and chitosan nanofibers: Electrospinning of chitin and deacetylation of chitin nanofibers. *Polymer,* **45**, pp. 7137–7142.

19. Taepaiboon, P., Rungsardthong, U., and Supaphol, P. (2006) Drug-loaded electrospun mats of poly(vinyl alcohol) fibers and their release characteristics of four model drugs. *Nanotechnology,* **17**, pp. 2317–2329.

20. Matthews, J.A., Wnek, G.E., Simpson, D.G., and Bowlin, G. (2002) Electrospinning of collagen nanofibers. *Biomacromolecules,* **3**, pp. 232–238.

21. Venugopal, J., Ma, L.L., Yong, T., and Ramakrishna, S. (2005) *In vitro* study of smooth muscle cells on polycaprolactone and collagen nanofibrous matrices. *Cell Boil. Int.,* **29**, pp. 861–867.

22. Rho, K.S., Jeong, L., Lee, G., Seo, B.M., Park, Y.J., Hong, S.D., and Min, B.M. (2006) Electrospinning of collagen nanofibers: Effects on the behavior of normal human keratinocytes and early-stage wound healing. *Biomaterials,* **27**, pp. 1452–1461.

23. Um, I.C., Fang, D.F., Hsiao, B.S., Okamoto, A., and Chu, B. (2004) Electro-spinning and electro-blowing of hyaluronic acid. *Biomacromolecules,* **5**, pp. 1428–1436.
24. Wang, X.F., Um, I.C., Fang, D.F., Okamoto, A., Hsiao, B.S., and Chu, B. (2005) Formation of water resistant hyaluronic acid nanofibers by blowing-assisted electro-spinning and non-toxic post treatments. *Polymer,* **46**, pp. 4853–4867.
25. Martins, A., Araújo, J.V., Reis, R.L., and Neves, N.M. (2007) Electrospun nanostructured scaffolds for tissue engineering applications. *Nanomedicine,* **2**, pp. 929–942.
26. Gilding, D.K., and Reed, A.M. (1979) Biodegradable polymers for use in surgery—polyglycolic/poly (lactic acid) homo- and copolymers: 1. *Polymer,* **20**, pp. 1459–1464.
27. Middleton, J.C., and Tipton, A.J. (2000) Synthetic biodegradable polymers as orthopedic devices. *Biomaterials,* **21**, pp. 2335–2346.
28. Barrows, T.H. (1986) Degradable implant materials: A review of synthetic absorbable polymers and their applications. *Clin. Mater.,* **1**, pp. 233–257.
29. Wang, E.A., Rosen, V., D'Alessandro, J.S., Bauduy, M., Cordes, P., Harada, T., Isreal, D.I., Hewick, R.M., Kerns, K.M., LaPan, P., Luxenberg, D., McQuaid, D., Moutsatsos, I.K., Nove, J., and Wozney, J.M. (1990) Recombinant human bone morphogenic protein induces bone formation. *Proc. Natl. Acad. Sci. USA,* **87**, pp. 2220–2224.
30. Ramchandani, M., and Robinson, D. (1998) In vitro release of ciprofloxacin from PLGA 50 : 50 implants. *J. Control. Release,* **54**, pp. 167–175.
31. Athanasiou, K.A., Agrawal, C.M., Barber, F.A., and Burkhart, S.S. (1998) Orthopaedic applications for PLA-PGA biodegradable polymers. *Arthroscopy,* **14**, pp. 726–737.
32. Oberpenning, F., Meng, J., Yoo, J.J., and Atala, A. (1999) De novo reconstitution of a functional mammalian urinary bladder by tissue engineering. *Nat. Biotechnol.,* **17**, pp. 149–155.
33. Park, K.I., Teng, Y.D., and Snyder, E.Y. (2002) The injured brain interacts reciprocally with neural stem cells supported by scaffolds to reconstitute lost tissue. *Nat. Biotechnol.,* **20**, pp. 1111–1117.
34. Kenawy, E.R., Layman, J.M., Watkins, J.R., Bowlin, G.L., Matthews, J.A., Simpson, D.G., and Wnek, G.E. (2003) Electrospinning of poly (ethylene-co-vinyl alcohol) fibers. *Biomaterials,* **24**, pp. 907–913.

35. Ryadnov, M.G., and Woolfson, D.N. (2003). Engineering the morphology of a self-assembling protein fibre. *Nat. Mater.*, **2**, pp. 329–332.

36. Frenot, A., and Chronakis, I.S. (2003) Polymer nanofibers assembled by electrospinning. *Curr. Opin. Colloid Interf.*, **8**, pp. 64–75.

37. Subbiah. T., Bhat, G.S., Tock, R.W., Parameswaran, S., and Ramkumar, S.S. (2005) Electrospinning of nanofibers. *J. Appl. Polym. Sci.*, **96**, pp. 557–569.

38. Hubbell, J.A. (1999) Bioactive biomaterials. *Curr. Opin. Biotech.*, **10**, pp. 123–129.

39. Griffith, L.G. (2002) Emerging design principles in biomaterials and scaffolds for tissue engineering. *Ann. NY. Acad. Sci.*, **961**, pp. 83–95.

40. Khor, E., and Lim, L.Y. (2003) Implantable applications of chitin and chitosan. *Biomaterials*, **24**, pp. 2339–2349.

41. Dvir, T., Tsur-Gang, O., and Cohen, S. (2005) "Designer" scaffolds for tissue engineering and regeneration. *Israel J. Chem.*, **45**, pp. 487–494.

42. Pierschbacher, M.D., and Ruoslahti, E. (1984) Cell attachment activity of fibronectin can be duplicated by small synthetic fragments of the molecule. *Nature*, **309**, pp. 30–33.

43. Li, M., Mondrinos, M.J., Gandhi, M.R., Ko, F.K., Weiss, A.S., and Lelkes, P.I. (2005) Electrospun protein fibers as matrices for tissue engineering. *Biomaterials*, **26**, pp. 5999–6008.

44. Li, J., He, A., Zheng, J., and Han, C.C. (2006) Gelatin and gelatin-hyaluronic acid nanofibrous membranes produced by electrospinning of their aqueous solutions. *Biomacromolecules*, **7**, pp. 2243–2247.

45. Li, C., Vepari, C., Jin. H.J., Kim, H.J., and Kaplan, D.L. (2006) Electrospun silk-BMP-2 scaffolds for bone tissue engineering. *Biomaterials*, **27**, pp. 3115–3124.

46. Bhardwaj, N., and Kundu, S.C. (2010) Electrospinning: A fascinating fiber fabrication technique. *Biotechnol. Adv.*, **28**, pp. 325–347.

47. Ma, Z., Kotaki, M., and Ramakrishna, S. (2005) Electrospun cellulose nanofiber as affinity membrane. *J. Membr. Sci.*, **265**, pp. 115–123.

48. Zhang, L., Menkhaus, T.J., and Fong, H. (2008) Fabrication and bioseparation studies of adsorptive membranes/felts made from electrospun cellulose acetate nanofibers. *J. Membr. Sci.*, **319**, pp. 176–184.

49. Duan, B., Dong, C., Yuan, X., and Yao, K. (2004). Electrospinning of chitosan solutions in acetic acid with poly (ethylene oxide). *J. Biomat., Sci. Polym. E*, **15**, pp. 797–811.

50. Chen, Z., Mo, X., and Qing, F. (2007) Electrospinning of collagen–chitosan complex. *Mater. Lett.,* **61**, pp. 3490–3494.

51. Wnek, G.E., Carr, M.E., Simpson, D.G., and Bowlin, G.L. (2003) Electrospinning of nanofibers fibrinogen structures. *Nano Lett.,* **3**, pp. 213–216.

52. Huang, Z.M., Zhang, Y.Z., Ramakrishna, S., and Lim, C.T. (2004) Electrospinning and mechanical characterization of gelatin nanofibers. *Polymer,* **45**, pp. 5361–5368.

53. Zarkoob, S., Eby, R.K., Reneker, D.H., Hudson, S.D., Ertley, D., and Adams, W.W. (2004). Structure and morphology of electrospun silk nanofibers. *Polymer,* **45**, pp. 3973–3977.

54. Jin, H.J., Fridrikh, S.V., Rutledge, G.C., and Kaplan, D.L. (2002) Electrospinning Bombyxmori silk with poly(ethylene oxide). *Biomacromolecules,* **3**, pp. 1233–1239.

55. Min, B.M., Lee, G., Kim, S.H., Nam, Y.S., Lee, T.S., and Park, W.H. (2004) Electrospinning of silk fibroin nanofibers and its effect on the adhesion and spreading of normal by human keratinocytes and fibroblasts *in vitro. Biomaterials,* **25**, pp. 1289–1297.

56. Min, B.M., Jeong, L., Nam, Y.S., Kim, J.M., Kim, J.Y., and Park, W.H. (2004). Formation of silk fibroin matrices with different texture and its cellular response to normal human keratinocytes. *Int. J. Biol. Macromol.,* **34**, pp. 223–230.

57. Kundu, S.C., Dash, B.C., Dash, R., and Kaplan, D.L. (2008) Natural protective glue protein, sericin bioengineered by silkworms: Potential for biomedical and biotechnological applications. *Prog. Polym. Sci.,* **33**, pp. 998–1012.

58. Takasu, Y., Yamada, H., and Tsubouchi, K. (2002) Isolation of three main sericin components from the cocoon of the silkworm, Bombyxmori. *Biosci. Biotechnol. Biochem.,* **66**, pp. 2715–2718.

59. McGrath, K., and Kaplan D. (1997). Silk, in *Protein-Based Materials* (Kaplan, D.L., Mello, C.M., Arcidiacono, S., Fossey, S., Senecal, K., and Muller, W., eds.), Birkhauser, Boston, pp. 103–131.

60. Altman, G.H., Horan, R.L., Lu, H.H., Moreau, J., Martin, I., Richmond, J.C., and Kaplan, D.L. (2002) Silk matrix for tissue engineered anterior cruciate ligaments. *Biomaterials,* **23**, pp. 4131–4141.

61. Dal-Pra, I., Freddi, G., Minic, J., Chiarini, A., and Armato, U. (2005) De novo engineering of reticular connective tissue *in vivo* by silk fibroin nonwoven materials. *Biomaterials,* **26**, pp. 1987–1999.

62. Meinel, L., Hofmann, S., Karageorgiou, V., Kirker-Head, C., McCool, J., Gronowicz, G., Zichner, L., Langer, R., Vunjak-Novakovic, G., and Kaplan, D.L. (2005) The inflammatory responses to silk films *in vitro* and *in vivo*. *Biomaterials*, **26**, pp. 147–155.
63. Wang, Y., Kim, H.J., Vunjak-Novakovic, G., and Kaplan, D.L. (2006) Stem cell-based tissue engineering with silk biomaterials. *Biomaterials*, **27**, pp. 6064–6082.
64. Minoura, N., Tsukada, M., and Nagura, M. (1990) Physico-chemical properties of silk fibroin membrane as a biomaterial. *Biomaterials*, **11**, pp. 430–434.
65. Wenk, E., Merkle, H.P., and Meinel, L. (2011) Silk fibroin as a vehicle for drug delivery applications. *J. Control. Release*, **15**, pp. 128–141.
66. Omenetto, F.G., and Kaplan, D.L. (2010) New opportunities for an ancient material, *Science*, **329**, pp. 528–531.
67. Das, S., Banerjee, R., and Bellare, J. (2005) Aspirin loaded albumin nanoparticles by coacervation: Implications in drug delivery. *Trends Biomater. Artif. Organs*, **18(2)**, pp. 203–212.
68. Hofmann, S., Foo, C.T., Rossetti, C.T.F., Textor, M., Vunjak-Novakovic, G., Kaplan, D.L., Merkle, H.P., and Meinel, L. (2006) Silk fibroin as an organic polymer for controlled drug delivery. *J. Control. Release*, **111(1–2)**, pp. 219–227.
69. Hardy, J.G., and Scheibel, T.R. (2010) Composite materials based on silk proteins. *Prog. Polym. Sci.*, **35(9)**, pp. 1093–1115.
70. Dyakonov, T., Yang, C.H., Bush, D., Gosangari, S., Majuru, S., and Fatmi, A. (2012) Design and characterization of a silk-fibroin-based drug delivery platform using naproxen as a model drug. *J. Drug Deliv.*, **2012**, pp. 10.
71. Mandal, B.B., Mann, J.K., and Kundu, S.C. (2009) Silk fibroin/gelatin multilayered films as a model system for controlled drug release. *Eur. J. Pharm. Sci.*, **37**, pp. 160–171.
72. Taheri, A., Atyabi, F., Nouri, F.S., Ahadi, F., Derakhshan, M.A., Amini, M., and Dinarvand, R. (2011) Nanoparticles of conjugated methotrexate-human serum albumin: Preparation and cytotoxicity evaluations. *J. Nanomaterials*, **2011**, pp. 7.
73. Cilurzo, F., Gennari, C.G., Selmin, F., Marotta, L.A., Minghetti, P., and Montanari, L. (2011) An investigation into silk fibroin conformation in composite materials intended for drug delivery. *Int. J. Pharm.*, **414**, pp. 218–224.

74. Mandal, B.B., Gil, E.S., Panilaitis, B., and Kaplan, D.L. (2013) Laminar silk scaffolds for aligned tissue fabrication. *Macromol. Biosci.*, **13**, pp. 48–58.

75. Queen, H. (2006) *Electrospinning Chitosan-Based Nanofibers for Biomedical Applications.* M.Sc. Thesis, North Carolina State University, Raleigh, NC, USA.

76. Di Martino, A., Sittinger, M., and Risbud, M.V. (2005) Chitosan: A versatile biopolymer for orthopaedic tissue-engineering. *Biomaterials*, **26**, pp. 5983–5990.

77. Paul, W., and Sharma, C.P. (2000) Chitosan, a drug carrier for the 21st century: A review. *STP Pharma Sci.*, **10**, pp. 5–22.

78. Muzzarelli, R.A.A., and Muzzarelli, C. (2005) Chitosan chemistry: Relevance to the biomedical sciences. *Adv. Polym. Sci.*, **186**, pp. 151–209.

79. Şenel, S., and McClure, S.J. (2004) Potential applications of chitosan in veterinary medicine. *Adv. Drug Deliv. Rev.*, **56**, pp. 1467–1480.

80. van der Rest, M., and Garrone, R. (1991) Collagen family of proteins. *FASEB J.*, **5**, pp. 2814–2823.

81. Badylak, S.F. (2002). The extracellular matrix as a scaffold for tissue reconstruction. *Semin. Cell Dev. Biol.*, **13**, pp. 377–383.

82. Dzemeshkevich, S.L., Konstantinov, B.A., Gromova, G.V., Lyudinovskova, R.A., and Kudrina, L.L. (1994) The mitral valve replacement by the new-type bioprostheses (features of design and long-term results). *J. Cardiovasc. Surg.*, **35**, pp. 189–191.

83. Klein, B., Schiffer, R., Hafemann, B., Klosterhalfen, B., and Zwadlo-Klarwasser, G. (2001) Inflammatory response to a porcine membrane composed of fibrous collagen and elastin as dermal substitute. *J. Mater. Sci. Mater. Med.*, **12**, pp. 419–424.

84. Zheng, M.H., Chen, J., Kirilak, Y., Willers, C., Xu, J., and Wood, D. (2005) Porcine small intestine submucosa (SIS) is not an acellular collagenous matrix and contains porcine DNA: Possible implications in human implantation. *J. Biomed. Mater. Res. B Appl. Biomater.*, **73**, pp. 61–67.

85. Domb, A.J., Kost, J., and Wiseman, M.W. (1997) Collagen: Characterization, processing, and medical applications, in *Handbook of Biodegradable Polymers* (Silver, F.H., and Garg, A.K., eds.), Harwood Academic Publishers, London, pp. 319–346.

86. O'Neill, P., and Booth, A.E. (1984) Use of porcine dermis as a dural substitute in 72 patients. *J. Neurosurg.*, **61**, pp. 351–354.

87. Allman, A.J., McPherson, T.B., Badylak, S.F., Merrill, L.C., Kallakury, B., Sheehan, C., Raeder, R.H., and Metzger, D.W. (2001) Xenogeneic extracellular matrix grafts elicit a TH2-restricted immune response. *Transplantation,* **71**, pp. 1631–1640.

88. Patino, M.G., Neiders, M.E., Andreana, S., Noble, B., and Cohen, R.E. (2003) Cellular inflammatory response to porcine collagen membranes. *J. Periodontal Res.,* **38**, pp. 458–464.

89. Konstantinovic, M.L., Lagae, P., Zheng, F., Verbeken, E.K., De Ridder, D., and Deprest, J.A. (2005) Comparison of host response to polypropylene and non-cross-linked porcine small intestine serosal-derived collagen implants in a rat model. *BJOG,* **112**, pp. 1554–1560.

90. Viswanadham, R.K., and Kramer, E.J. (1976) Elastic properties of reconstituted collagen hollow fiber membranes. *J. Mater. Sci.,* **11**, pp. 1254–1262.

91. Dougados, M., Nguyen, M., Listrat, V., and Amor, B. (1993) High molecular weight sodium hyaluronate (hyalectin) in osteoarthritis of the knee: A 1 year placebo-controlled trial. *Osteoarth. Cartilage,* **1**, pp. 97–103.

92. Krause, W., Bellamo, E., and Colby, R. (2001) Rheology of sodium hyaluronate under physiological conditions. *Biomacromolecules,* **2**, pp. 65–69.

93. Li, W.J., Laurencin, C.T., Caterson, E.J., Tuan, R.S., and Ko, F.K. (2002) Electrospun nanofibrous structure: A novel scaffold for tissue engineering. *J. Biomed. Mater. Res.,* **60**, pp. 613–621.

94. Mark, H.F., Bikales, N.M., Overberger, C.G., Menges, G., and Kroschwitz, J.I. (1989) *Encyclopedia of Polymer Science and Engineering,* 2nd ed. (Wiley, New York, USA).

95. Ma, H., Zeng, J., Realff, M.L., Kumar, S., and Schiraldi, D.A. (2003) Processing, structure, and properties of fibers from polyester/carbon nanofiber composites. *Compos. Sci. Technol.,* **63**, pp. 1617–1628.

96. Ward, A.G., and Courts, A. (1977) *The Science and Technology of Gelatin*, Illustrated edition (Academic Press, London).

97. Choi, Y.S., Hong, S.R., Lee, Y.M., Song, K.W., Park, M.H., and Nam, Y.S. (1999) Study on gelatin-containing artificial skin: I. Preparation and characteristics of novel gelatin–alginate sponge. *Biomaterials,* **20**, pp. 409–417.

98. Ulubayram, K., Cakar, A.N., Korkusuz, P., Ertan, C., and Hasirci, N. (2001) EGF containing gelatin-based wound dressings. *Biomaterials,* **22**, pp. 1345–1356.

99. Tabata, Y., Hijikata, S., and Ikada, Y. (1994) Enhanced vascularization and tissue granulation by basic fibroblast growth factor impregnated in gelatin hydrogels. *J. Control. Release,* **31**, pp. 189–199.

100. Cortesi, R., Nastruzzi, C., and Davis, S.S. (1998) Sugar cross-linked gelatin for controlled release: Microspheres and disks. *Biomaterials,* **19**, pp. 1641–1649.

101. Li, J.K., Wang, N., and Wu, X.S. (1998) Gelatin nanoencapsulation of protein/peptide drugs using an emulsifier-free emulsion method. *J. Microencapsul.,* **15**, pp. 163–172.

102. Guidoin, R., Marceau, D., Rao, T.J., King, M., Merhi, Y., Roy, P.E., Martin, L., and Duval, M. (1987) *In vitro* and *in vivo* characterization of an impervious polyester arterial prosthesis: The Gelseal Triaxial graft. *Biomaterials,* **8**, pp. 433–441.

103. Jonas, R.A., Ziemer, G., Schoen, F.J., Britton, L., and Castaneda, A.R. (1988) A new sealant for knitted Dacron prostheses minimally crosslinked gelatin. *J. Vasc. Surg.,* **7**, pp. 414–419.

104. Marois, Y., Chakfe, N.L., Deng, X., Marois, M., How, T., King, M.W., and Guidoin, R. (1995) Carbodiimide cross-linked gelatin: A new coating for porous polyester arterial prostheses. *Biomaterials,* **16**, pp. 1131–1139.

105. Nagura, M., Yokota, H., Ikeura, M., Gotoh, Y., and Ohkoshi, Y. (2002) Structures and physical properties of cross-linked gelatin fibers. *Polym. J.,* **34**, pp. 761–766.

106. Giusti, P., Barbani, N., Lazzeri, L., Polacco, G., Cristallini, C., and Cascone, M.G. (1998) Gelatin-poly(vinyl alcohol) blends as bioartificial polymeric materials. *Proc. 4th Int. Conference on Frontiers Polym. Adv. Mater.,* ICFPAM., pp. 449–462 (in English).

107. Mosesson, M.W. (2005) Fibrinogen and fibrin structure and functions. *J. Thromb. Haemost.,* **3**, pp. 1894–1904.

108. Gugutkov, D., Gustavsson, J., Ginebra, M.P., and Altankov, G. (2013) Fibrinogen nanofibers for guiding endothelial cell behavior. *Biomat. Sci.,* **1**, pp. 1065–1073.

109. Lisman, T., Weeterings, C., and de Groot, P.G. (2005) Platelet aggregation: Involvement of thrombin and fibrin (ogen). *Front Biosci.,* **10**, pp. 2504–2517.

110. Weisel, J.W. (2005) Fibrinogen and fibrin. *Adv. Protein Chem.,* **70**, pp. 247–299.

111. Raghow, R. (1994) The role of extracellular matrix in postinflammatory wound healing and fibrosis. *FASEB J.,* **8**, pp. 823–831.

112. Clark, R.A.F. (1996) *The Molecular and Cellular Biology of Wound Repair,* 2nd edn. (Plenum Press, New York).

113. Flick, M.J., Du, X., Witte, D.P., Jirouskova, M., Soloviev, D.A., Busuttil, S.J., Plow, E.F., and Degen, J.L. (2004) Leukocyte engagement of fibrin(ogen) via the integrin receptor alphambeta2/mac-1 is critical for host inflammatory response *in vivo. J. Clin. Invest.,* **113**, pp. 1596–1606.

114. Vanhinsbergh, V.W., Collen, A., and Koolwijk, P. (2001) Role of fibrin matrix in angiogenesis. *Ann. NY Acad. Sci.,* **936**, pp. 426–437.

115. Dietrich, M., Heselhaus, J., Wozniak, J., Weinandy, S., Mela, P., Tschoeke, B., Schmitz-Rode, T., and Joeckenhoevel, S. (2013) Fibrin-based tissue engineering: Comparison of different methods of autologous fibrinogen isolation. *Tissue Eng., Part C,* **3**, pp. 216–226.

116. Hojo, M., Inokuchi, S., Kidokoro, M., Fukuyama, N., Tanaka, E., Tsuji, C., Miyasaka, M., Tanino, R., and Nakazawa, H. (2003) Induction of vascular endothelial growth factor by fibrin as a dermal substrate for cultured skin substitute. *Plast. Reconstr. Surg.,* **111**, pp. 1638–1645.

117. Hunter, C.J., Mouw, J.K., and Levenston, M.E. (2004) Dynamic compression of chondrocyte-seeded fibrin gels: Effects on matrix accumulation and mechanical stiffness. *Osteoarthritis Cartilage,* **12**, pp. 117–130.

118. Davis, H.E., Miller, S.L., Case, E.M., and Leach, J.K. (2011) Supplementation of fibrin gels with sodium chloride enhances physical properties and ensuing osteogenic response. *Acta Biomaterialia,* **7**, pp. 691–699.

119. Mcmanus, M.C., Boland, E.D., Simpson, D.G., Barnes, C.P., and Bowlin, G.L. (2007) Electrospun fibrinogen: Feasibility as a tissue engineering scaffold in a rat cell culture model. *J. Biomed. Mater. Res. A.,* **81**, pp. 299–309.

120. Frazza, E.J., and Schmitt, E.E. (1971) A new absorbable suture. *J. Biomed. Mater. Res.,* **1**, pp. 43–58.

121. Conn Jr, J., Oyasu, R., Welsh, M., and Beal, J.M. (1974) Vicryl (polyglactin 910) synthetic absorbable sutures. *Am. J. Surg.,* **128**, pp. 19–23.

122. Ray, J.A., Doddi, N., Regula, D., Williams, J.A., and Melveger, A. (1981) Polydioxanone (PDS), a novel monofilament synthetic absorbable suture. *Surg. Gynecol. Obstet.,* **153**, pp. 497–507.

123. Katz, A.R., Mukherjee, D.I., Kaganov, A.L., and Gordon, S. (1985) A new synthetic monofilament absorbable suture made from polytrimethylene carbonate. *Surg. Gynecol. Obstet.,* **161**, pp. 213–222.

124. Bezwada, R.S., Jamiolkowski, D.D., Lee, I.Y. Agarwal, V., Persivale, J., Trenka-Benthin, S., Erneta, M., Suryadevara, J., Yang, A., and Liu, S. (1995) Monocryl@ suture, a new ultra-pliable absorbable monofilament suture. *Biomaterials,* **16**, pp. 1141–1148.

125. Leung, K.S. (1994) Bioresorbable internal fixation devices: Mechanical properties and future trends in production technologies, in *Biodegradable Implants in Fracture Fixation* (Gogolewski, S., ed.), Chinese University Hong Kong & World Scientific Publishing, Hong Kong, pp. 249–258.

126. Gregoriadis, G. (1979) Lactic/glycolic acid polymers, in *Drug Carriers in Biology and Medicine* (Wise, D.L., Fellmann, T.D., Sanderson, J.E., and Wentworth, R.L., eds.), Academic Press, London, pp. 237–270.

127. Rajasubramanian, G., Meidell, R.S., Landau, C., Dollar, M.L., Holt, D.B., Willard, J.E., Prager, M.D., and Eberhart, R.C. (1994) Fabrication of resorbable microporous intravascular stents for gene therapy applications. *ASAIO J.,* **40**, pp. 584–589.

128. Freed, L.E., Grande, D.A., Lingbin, Z., Emmanual, J., Marquis, J.C., and Langer, R. (1994) Joint resurfacing using allograft chondrocytes and synthetic biodegradable polymer scaffolds. *J. Biomed. Mater. Res.,* **28**, pp. 891–899.

129. Gugala, Z., and Gogolewski, S. (2000) *In vitro* growth and activity of primary chondrocytes on a resorbable polylactide three dimensional scaffold. *J. Biomed. Mater. Res.,* **49**, pp. 183–191.

130. Pineda, L.M., Busing, M.C., Meinig, R.P., and Gogolewski, S. (1996) Bone regeneration with resorbable polymeric membrane. III. Effect of poly (L-lactide) membrane pore size on the bone healing process in large defects. *J. Biomed. Mater. Res.,* **31**, pp. 385–394.

131. Gugala, Z., and Gogolewski, S. (1999) Regeneration of segmental diaphyseal defects in the sheep tibiae using resorbable polymeric membranes. A preliminary study. *J. Orthop. Trauma.,* **13**, pp. 187–195.

132. Ratner, B.D., Hoffman, A.S., Schoen, F.J., and Lemons, J.E. (1996) Bioresorbable and bioerodible materials, in *Biomaterials Science* (Kohn, J., and Langer, R., eds.), Academic Press, New York, pp. 64–72.

133. Shalaby, S.W. (1994) *Biomedical Polymers: Designed to Degrade Systems.* Illustrated edition (Hanser Publishers, New York).

134. Daniels, A.U., Chang, M.K.O., Andriano, K.P., and Heller, J. (1990) Mechanical properties of biodegradable polymers and composites proposed for internal fixation of bone. *J. Appl. Biomater.,* **1**, pp. 57–78.

135. Wise, D.L., Trantolo, D.J., Altobelli, D.E., Yaszemski, M.J., Greser, J.D., and Schwartz, E.R. (1995) The use of PLA-PGA polymers in orthopedics, in *Encyclopedic Handbook of Biomaterials and Bioengineering. Part A—Materials* (Agrawal, C.M., Niederauer, G.G., Micallef, D.M., and Athanasiou, K.A., eds.), Marcel Dekker, New York, pp. 1055–1089.

136. Bergsma, J.E., de Bruijn, W.C., Rozema, F.R., Bos, R.R.M., and Boering, G. (1995) Late degradation tissue response to poly(L-lactide) bone plates and screws. *Biomaterials*, **16**, pp. 25–31.

137. Schindler, A., Jeffcoat, R., Kimmel, G.L., Pitt, C.G., Wall, M.E., and Zwiedinger, R. (1977) Biodegradable polymers for sustained drug delivery. *Contemp. Top. Polym. Sci.*, **2**, pp. 251–289.

138. Barber, F. A. (1998) Resorbable fixation devices: A product guide. *Orthopedic Special Edition*, **4**, pp. 1111–1117.

139. Myer, K. (2009) *Biomedical Engineering and Design Handbook*, 2nd edn. (The McGraw-Hill companies, New York).

140. Michaeli, W., and von Oepen, R. (1994) Processing of degradable polymers. *ANTEC*, pp. 796–804.

141. von Oepen, R., and Michaeli, W. (1992) Injection moulding of biodegradable implants. *Clin. Mater.*, **10**, pp. 21–28.

142. Middleton, J.C., Williams, C.T., and Swaim, R.P. (1997) The melt viscosity of polyglycolic acid as a function of shear rate, moisture, and inherent viscosity. *Trans. Soc. Biomater. 23rd Annu. Meeting*, **20**, pp. 106.

143. Formhals, A. (1934) Process and apparatus for preparing artificial threads. U.S. Patent No. 1, 975, **504.**

144. Li, D., and Xia, Y. (2004) Electrospinning of nanofibers: Reinventing the wheel? *Adv. Mater.*, **16**, pp. 1151–1170.

145. Venugopal, J., and Ramakrishna, S. (2005) Applications of polymer nanofibers in biomedicine and biotechnology. *Appl. Biochem. Biotechnol.*, **125**, pp. 147–158.

146. Ma, Z., Kotaki, M., Inai, R., and Ramakrishna, S. (2005) Potential of nanofiber matrix as tissue-engineering scaffolds. *Tissue Eng.*, **11**, pp. 101–109.

147. Wu, X.H., Wang, L.G., and Huang, Y. (2006) Application of electrospun ethyl cellulose fibers in drug release systems. *Acta Polymerica Sinica*, **2**, pp. 264–268.

148. Schindler, M., Ahmed, I., Kamal, J., Nur-E-Kamal, A., Grafe, T.H., Young Chung, H., and Meiners. S. (2005) A synthetic nanofibrillar matrix promotes *in vivo*-like organization and morphogenesis for cells in culture. *Biomaterials*, **26**, pp. 5624–5631.

149. Fridrikh, S.V., Yu, J.H., Brenner, M.P., and Rutledge, G.C. (2003) Controlling the fiber diameter during electrospinning. *Phys. Rev. Lett.,* **90**, pp. 144502–144502.

150. Mit-uppatham, C., Nithitanakul, M., and Supaphol, P. (2004) Ultrafine electrospun polyamide-6 fibers: Effect of solution conditions on morphology and average fiber diameter. *Macromol. Chem. Phys.,* **205**, pp. 2327–2338.

151. Dees, J.R., and Spruiell, J.E. (1974) Structure development during melt spinning of linear polyethylene fibers. *J. Appl. Polym. Sci.,* **18**, pp. 1053–1078.

152. Ratner, B.D., Hoffman, A.S., Schoen, F.J., and Lemons, J.E. (1997) *Biomaterials Science: An Introduction to Materials in Medicine,* 1st edn. (Academic Press, New York).

153. Ayutsede, J., Gandhi, M., Sukigara, S., Micklus, M., Chen, H.E., and Ko, F. (2005) Regeneration of Bombyxmori silk by electrospinning. Part 3: Characterization of electrospun nonwoven mat. *Polymer,* **46**, pp. 1625–1634.

154. Zussman, E., Yarin, A., and Weihs, D. (2002) A micro-aerodynamic decelerator based on permeable surfaces of nanofiber mats. *Exp. Fluids,* **33**, pp. 315–320.

155. Huang, L., McMillan, R.A., Apkarian, R.P., Pourdeyhimi, B., Conticello, V.P., and Chaikof, E.L. (2000) Generation of synthetic elastin-mimetic small diameter fibers and fiber networks. *Macromolecules,* **33**, pp. 2989–2997.

156. Tan, E.P.S., Ng, S.Y., and Lim, C.T. (2005) Tensile testing of a single ultrafine polymeric fiber. *Biomaterials,* **26**, pp. 1453–1456.

157. Haque, M.A., and Saif, M.T.A. (2003) A review of MEMS-based microscale and nanoscale tensile and bending testing. *Exp. Mech.,* **43**, pp. 248–255.

158. DiNardo, N.J. (1994) *Nanoscale Characterization of Surfaces and Interfaces.* 163S VCH (Federal Republic of Germany).

159. Ramanathan, K., Bangar, M.A., Yun, M., Chen, W., Myung, N.V., and Mulchandani, A. (2005) Bioaffinity sensing using biologically functionalized conducting-polymer nanowire. *J. Am. Chem. Soc.,* **127**, pp. 496–497.

160. Wenger, M.P.E., Bozec, L., Horton, M.A., and Mesquida, P. (2007) Mechanical properties of collagen fibrils. *Biophys. J.,* **93**, pp. 1255–1263.

161. Sun, L., Han, R.P.S., Wang, J., and Lim, C.T. (2008) Modeling the size-dependent elastic properties of polymeric nanofibers. *Nanotechnology*, **19**, pp. 1–8.
162. Coburn, J.C., and Pandit, A. (2007) Development of naturally-derived biomaterials and optimization of their biomechanical properties. *Top. Tissue Eng.*, **3**, pp. 1–14.
163. Dickerson, M.T. (2012) *Protein Based Biomimetic Approaches to Surface Hemocompatibility and Biocompatibility Enhancement.* Thesis and Dessertations-Chemical and Materials Engineering. Paper 6. http://uknowledge.uky.edu/cme_etds/6
164. Waldman, S.D., and Lee, J.M. (2005) Effect of sample geometry on the apparent biaxial mechanical behaviour of planar connective tissues. *Biomaterials*, **26**, pp. 7504–7513.
165. Billiar, K.L., and Sacks, M.S. (2000) Biaxial mechanical properties of the natural and glutaraldehyde treated aortic valve cusp—Part I: Experimental results. *J. Biomech. Eng.*, **122**, pp. 23–30.
166. Font, P.J., Del, O.B.M., Castro, F.M.B., Infante, M.A., Alonso, V.A.I., and Palomares, C.T. (2013) *New Biomaterial Based on Whorton's Jelly from the Human Umbilical Cord.* European Patent No. EP 2351538. (European Patent Office, Munich, Germany).
167. Choi, H.S., and Vito, R.P. (1990) Two-dimensional stress-strain relationship for canine pericardium. *J. Biomech. Eng.*, **112**, pp. 153–159.
168. Wasserman, A.J., Doillon, C.J., Glasgold, A.I., Kato, Y.P., Christiansen, D., Rizvi, A., Wong, E., Goldstein, J., and Silver, F.H. (1988) Clinical applications of electron microscopy in the analysis of collagenous biomaterials. *Scanning Microsc.*, **2**, pp. 1635–1646.
169. Fackler, K., Klein, L., and Hiltner, A. (1981) Polarizing light microscopy of intestine and its relationship to mechanical behaviour. *J. Microsc.*, **124**, pp. 305–311.
170. Kronick, P.L., and Buechler, P.R. (1986) Fiber orientation in calf skin by laser light scattering or X-ray diffraction and quantitative relation to mechanical properties. *J. Am. Leather. Chem. Assoc.*, **81**, pp. 221–229.
171. Kronick, P.L., and Sacks, M.S. (1991) Quantification of vertical-fiber defect in cattle hide by small-angle light scattering. *Connect Tissue Res.*, **27**, pp. 1–13.
172. Hilbert, S.L., Sword, L.C., Batchelder, K.F., Barrick, M.K., and Ferrans, V.J. (1996) Simultaneous assessment of bioprosthetic heart valve

biomechanical properties and collagen crimp length. *J. Biomed. Mater. Res.,* **31**, pp. 503–509.

173. Geday, M.A., Kaminsky, W., Lewis, J.G., and Glazer, A.M. (2000) Images of absolute retardance L.Deltan, using the rotating polariser method. *J. Microsc.,* **198**, pp. 1–9.

174. Tower, T.T., Neidert, M.R., and Tranquillo, R.T. (2002) Fiber alignment imaging during mechanical testing of soft tissues. *Ann. Biomed. Eng.,* **30**, pp. 1221–1233.

175. Huang, Z.M., Zhang, Y.Z., Kotaki, M., and Ramakrishna, S. (2003) A review on polymer nanofibers by electrospinning and their applications in nanocomposites. *Compos. Sci. Technol.,* **63**, pp. 2223–2253.

176. Kenawy, El-R., Bowlin, G.L., Mansfield, K., Layman, J., Simpson, D.G., Sanders, E.H., and Wnek, G.E. (2002) Release of tetracycline hydrochloride from electrospun poly (ethylene-co-vinylacetate), poly (lactic acid), and a blend. *J. Control. Release,* **81**, pp. 57–64.

177. Ratner, B.D. (1993) New ideas in biomaterials science: a path to engineered biomaterials. *J. Biomed. Mater. Res.,* **27**, pp. 837–850.

178. Katti, D.S., Robinson, K.W., Ko, F.K., and Laurencin, C.T. (2004) Bioresorbable nanofiber-based systems for wound healing and drug delivery: Optimization of fabrication parameters. *J. Biomed. Mater. Res. B*: *Appl. Biomater.,* **70**, pp. 286–296.

179. Jiang, H., Fang, D., Hsiao, B., Chu, B., and Chen, W. (2004) Preparation and characterization of ibuprofen-loaded poly(lactide-co-glycolide)/poly(ethylene glycol)-g-chitosan electrospun membranes. *J. Biomater. Sci. Polym.,* **15**, pp. 279–296.

180. Zeng, J., Xu, X., Chen, X., Liang, Q., Bian, X., Yang, L., and Jing, X. (2003) Biodegradable electrospun fibers for drug delivery. *J. Control. Release,* **92**, pp. 227–231.

181. Verreck, G., Chun, I., Peeters, J., Rosenblatt, J., and Brewster, M.E. (2003) Preparation and characterization of nanofibers containing amorphous drug dispersions generated by electrostatic spinning. *Pharm. Res.,* **20**, pp. 810–817.

182. Kim, K., Luu, Y.K., Chang, C., Fang, D., Hsiao, B.S., Chu, B., and Hadjiargyrou, M. (2004) Incorporation and controlled release of a hydrophilic antibiotic using poly (lactide-co-glycolide)-based electrospunnanofibrous scaffolds. *J. Control. Release,* **98**, pp. 47–56.

183. Zhang, Y.Z., Venugopal, J., Huang, Z.M., Lim, C.T., and Ramakrishna, S. (2006) Crosslinking of the electrospun gelatin nanofibers. *Polymer,* **47**, pp. 2911–2917.

184. Dai, N.T., Williamson, M.R., Khammo, N., Adams, E.F., and Coombes, A.G.A. (2004) Composite cell support membranes based on collagen and polycaprolactone for tissue engineering of skin. *Biomaterials,* **25**, pp. 4263-4271.
185. Yannas, I.V. (1998). Studies on the biological activity of the dermal regeneration template. *Wound Repair Regen.*, **6**, pp. 518-524.
186. Venugopal, J., and Ramakrishna, S. (2005) Biocompatible nanofiber matrices for the engineering of a dermal substitute for skin regeneration. *Tissue Eng.,* **11**, pp. 847-854.
187. Friess, W. (1998) Collagen—biomaterial for drug delivery. *Eur. J. Pharm. Biopharm.,* **45**, 113-136.
188. Almany, L., and Seliktar, D. (2005) Biosynthetic hydrogel scaffolds made from fibrinogen and polyethylene glycol for 3d cell cultures. *Biomaterials,* **26**, pp. 2467-2477.
189. Yoo, H.S., Lee, E.A., Yoon, J.J., and Park, T.G. (2005) Hyaluronic acid modified biodegradable scaffolds for cartilage tissue engineering. *Biomaterials,* **26**, pp. 1925-1933.
190. Mo, X.M., Xu, C.Y., Kotaki, M., and Ramakrishna, S. (2004) Electrospun P (LLA-CL) nanofiber: A biomimetic extracellular matrix for smooth muscle cell and endothelial cell proliferation. *Biomaterials,* **25**, pp. 1883-1890.
191. Zong, X., Bien, H., Chung, C.Y., Yin, L., Fang, D., Hsiao, B.S., and Entcheva, E. (2005) Electrospun fine-textured scaffolds for heart tissue constructs. *Biomaterials,* **26**, pp. 5330-5338.
192. Yang, F., Murugan, R., Ramakrishna, S., Wang, X., Ma, Y.X., and Wang, S. (2004) Fabrication of nanostructured porous PLLA scaffold intended for nerve tissue engineering. *Biomaterials,* **25**, pp. 1891-1900.
193. Kobayashi, H., Yokoyama, Y., Takato, T., Koyama, H., and Ichioka, S. (2007). *Spongiform Structured Materials and its Manufacturing Methods.* Japanese patent application number 2007-103201.
194. Shum-Tim, D., Stock, U., Hrkach, J., Shinoka, T., Lien, J., Moses, M.A., and Mayer Jr, J.E. (1999) Tissue engineering of autologous aorta using a new biodegradable polymer. *Ann. Thorac. Surg.,* **68**, pp. 2298-2304.
195. Niklason, L.E., Gao, J., Abbott, W.M., Hirschi, K.K., Houser, S., Marini, R., and Langer, R. (1999) Functional arteries grown *in vitro. Science,* **284**, pp. 489-493.
196. Wu, X., Rabkin-Aikawa, E., Guleserian, K.J., Perry, T.E., Masuda, Y., Sutherland, F.W., and Bischoff, J. (2004) Tissue-engineered microvessels on three-dimensional biodegradable scaffolds using

human endothelial progenitor cells. *Am. J. Physiol. Heart C.,* **287**, pp. 480–487.

197. Gao, J., Ensley, A.E., Nerem, R.M., and Wang, Y. (2007) Poly (glycerol sebacate) supports the proliferation and phenotypic protein expression of primary baboon vascular cells. *J. Biomed. Mater. Res. A.,* **83**, pp. 1070–1075.

198. Xu, Z.C., Zhang, W.J., Li, H., Cui, L., Cen, L., Zhou, G.D., Liu, W., and Cao, Y. (2008) Engineering of an elastic large muscular vessel wall with pulsatile stimulation in bioreactor, *Biomaterials,* **29**, pp. 1464–1472.

199. Freytes, D.O., Rundell A.E., Vande Geest, J., Vorp, D.A., Webster, T.J., Badylak, S.F. (2005) Analytically derived material properties of multilaminated extracellular matrix devices using the ball-burst test. *Biomaterials,* **26**(27), 5518–5531.

Chapter 5

Biomaterials in Total Hip Joint Replacements: The Evolution of Basic Concepts, Trends, and Current Limitations—A Review

B. Bhaskar,[a] S. Arun,[b] P. S. Rama Sreekanth,[c] and S. Kanagaraj[b]

[a]*Department of Orthopaedics, North Eastern Indira Gandhi Regional Institute of Health and Medical Sciences, Shillong-793018, Meghalaya, India*
[b]*Department of Mechanical Engineering, Indian Institute of Technology Guwahati, Guwahati-781039, Assam, India*
[c]*Department of Mechanical Engineering, National Institute of Science and Technology, Berhampur-761008, Odisha, India*
kanagaraj@iitg.ernet.in

In the successful clinical results of total hip replacement (THR), technological developments in orthopedic implants have been extremely important. THR replaces an unsalvageable joint to restore its normal function by providing a painless, mobile, and stable joint. Several progresses have been made to increase long-term results through the incorporation of newer biomaterials, improved articulating surface design to reduce wear, and fixation techniques.

Trends in Biomaterials
Edited by G. P. Kothiyal and A. Srinivasan
Copyright © 2016 Pan Stanford Publishing Pte. Ltd.
ISBN 978-981-4613-98-9 (Hardcover), 978-981-4613-99-6 (eBook)
www.panstanford.com

This chapter aims to show not only the accelerated developments and significant advances in biomaterials used in THR but also several therapeutic risks that should not be ignored. Thus, the emerging risk–benefit scenario of doing a THR and its evolution thereof are discussed in more detail. The biomechanics and stability of implants and their role in implant longevity (survivorship) have also been described with a glimpse of merit–demerit analysis of several biomaterials used as the articulating surfaces in THR. This chapter focusses on the fixation methods and THR implant materials to understand the common causes of failures and subsequent revision surgeries and possible solutions for reducing the same. The technical problems faced by surgeons during implant fixation in THR and in the selection of materials based on the patient's requirement are also discussed in detail. The recent advances in implant materials, including the reinforcement of ultrahigh molecular weight polyethylene (UHMWPE) and polymethyl methacrylate (PMMA) bone cement by multiwalled carbon nanotubes (MWCNTs), are also discussed. A brief discussion on biocompatibility issues related to THR has also been deliberated.

5.1 Introduction

A biomaterial is defined as "a nonviable material used in a medical device intended to interact with biological systems." THR is performed to relieve pain and improve activities of patients. Total hip arthroplasty (THA) aims to provide a painless, mobile, stable, and durable joint. This goal was attained by replacing the unsalvageable joint to restore its normal function. The newly implanted joints need to be sturdy and able to bear the body weight, providing the required stability and strength. Although biomaterial development has been booming with new information and the scope of THA is expanding rapidly, the flow of information across the global community is rather slow. The information available in open resources can be used to enhance educational experience as well as to create awareness about health care [1]. Our knowledge about research, education, and clinical affairs depends on understanding the reasons behind THA by biologists, engineers, and orthopedic surgeons based on the published literature and expertise available to them. The literature survey not only showed accelerated developments and significant advances in biomaterials used in THR, it also reported several avoidable therapeutic risks to increase the lifespan of the implant.

5.2 Components of THR

The typical components of THR prosthesis are shown in Fig. 5.1 [2]. The primary components of an artificial hip consist of (1) an acetabular cup typically made of metal, UHMWPE, or ceramic; (2) the ball or head, which replaces the head of the femur, typically made of cobalt–chromium (Co–Cr) alloy, stainless steel, or ceramic materials such as aluminum oxide or zirconium oxide; and (3) the stem, which fits into the femur, typically made of Co–Cr alloy, titanium alloy (Ti6Al4V), or rarely 316L stainless steel [3–5]. The components of THR must provide the critical movements for the required functions, and they should be able to function inside the body without causing any immune system reaction. Moreover, the results of such surgeries must be predictable and reproducible.

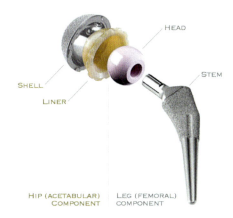

Figure 5.1 Components of artificial hip prosthesis [2].

The primary reasons for which THR is carried out in Western countries are osteoarthritis (93%), avascular necrosis (2%), fractured neck of femur (2%), congenital dislocation (1%), and inflammatory arthropathy (abnormality of a joint) (1%) [6, 7]. Figure 5.2 [7] gives statistics of various reasons leading to both primary and revision hip surgery. The THR performed on a patient is not once-in-a-life-time procedure; the necessity for revised surgeries depends on several surgical and patient-related factors such as the activity level and age of the patient, expertise of the surgeon, and

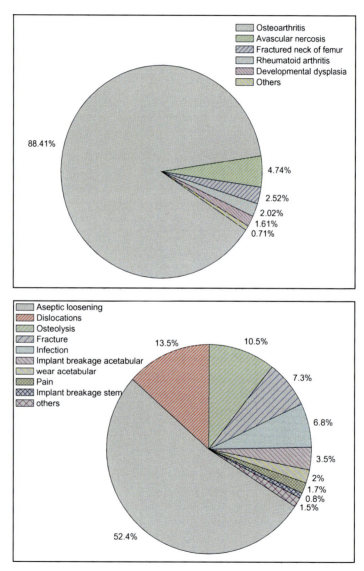

Figure 5.2 Statistics showing various factors leading to (a) primary and (b) revision of hip surgeries [7].

postoperative care. The prominent reasons for revised surgeries have been identified as aseptic loosening (52.4%), dislocation of prosthesis (13.5%), osteolysis (10.5%), fracture (7.3%), and wear of acetabular (2%) [7]. To understand the current practices and limitations of THA, having a glimpse of its evolution is imperative. The next section discusses the evolution of THA technology down the timeline.

5.3 Evolution of Hip Arthroplasty Technology

Hip arthroplasty has a history spanning more than 100 years. From a historical perspective, German Professor Themistocles Glück [8] (1853–1942) led the way in the development of hip implant fixation. In 1891, Glück fixed an ivory ball-and-socket joint in the bone with nickel-plated screws. Subsequently, he experimented with a mixture of plaster of Paris and powdered pumice with resin to provide fixation [9]. Later during the 1940s, Austin Moore [10] from the Johns Hopkins, USA tried to use Vitallium (cobalt–chrome alloy) for joint replacement. The first metal prosthesis was used at Johns Hopkins Hospital in 1940. In France, the Judet brothers [11] used an acrylic prosthesis in 1948 [12]. It was refined by Frederick Röeck Thompson (1907–83) [13], who developed Vitallium prosthesis in 1950, which featured a distinctive flared collar below the head and a vertical intramedullary stem. Later in 1952, Moore and Böhlman [13] refined their implant, which featured a stem to allow bone ingrowth. These first hip arthroplasty products were widely distributed and used for the replacement of femoral head and neck. In 1962, Charnley [14] observed that the "hard-on-hard" bearing surfaces, like the metal-on-metal components, offered very high frictional torque in the bearing metallic surfaces as the technology for manufacturing highly polished metallic bearing surfaces was not available. Charnley found that UHMWPE was more biocompatible and had adequate bearing surface compared to Teflon due to its low coefficient of friction with metallic bearing surfaces. This heralded a "hard-on-soft" combination for *low friction torque arthroplasty* [13, 15]. In 1964, Briton started the clinical experience with metal-on-metal articulation along with cementless fixation. Some of these early arthroplasties provided surprisingly good results with up to 97% of implants surviving more than 17 years of follow-up [13]. In 1970, Boutin developed ceramic-on-ceramic hip arthroplasty. However, in favor of Charnley's cemented hard-on-soft model, most of the hard-on-hard designs were abandoned [15]. The basic concept of a cemented UHMWPE cup and metallic stem is still a gold standard in THA. Figure 5.3 discusses the evolution of THA over timelines discussed above. In spite of initial success, problems due to aseptic loosening and wear debris from the polyethylene component due to abrasive, adhesive, and third-body wear emerged as new setbacks for Charnley's design [15]. Despite operational shortcomings, metal-on-polyethylene is still favored by most orthopedicians compared to other combinations of articulating surfaces.

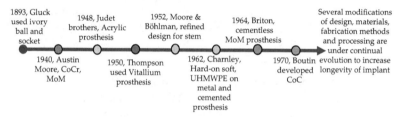

Figure 5.3 Timeline of evolution of THA.

5.4 Currently Used Biomaterials in Bearing Surfaces

The long-term success of THA lies in the robustness of the bearing surface to withstand impact loading and wear. In general, there are two options for bearing surfaces in THR: hard-on-soft and hard-on-hard; each one of them has two sub-classifications as follows:

5.4.1 Hard-on-Soft Combinations

1. *Metal-on-Polyethylene*: The ball and the acetabular liner are made of metal and polyethylene, respectively. Metal-on-polyethylene (MoP) bearings, which were popularized by Charnley in the 1970s, are the extensively used and followed prosthesis of the current day in THA. The specific advantage of MoP is to provide a cost-effective bearing with predictability and safety for a majority of patients. Thus, MoP is considered

as a gold standard in THA [16]. At present, most of the other bearing surfaces have been replaced by polyethylene-based implants, leading to a large proportion of research aimed at developing the design and improving the material properties as well as implantation techniques to further enhance the longevity of the MoP implant prosthesis [13, 17]. Figure 5.4 shows an MoP implant, without bone cement.

2. *Ceramic-on-polyethylene*: The ball and the acetabular liner are made of ceramic and polyethylene, respectively.

Figure 5.4   Metal-on-polyethylene articulating surface [18].

Ceramic bearing surfaces offer the highest wear resistance compared to polyethylene. They provide long-term mechanical reliability and offer high resistance to deformation, high fracture toughness, and high fatigue resistance, which are vital for materials used as articulating surfaces in total joint arthroplasty. Nonetheless, ceramic bearings are not popular among hip surgeons due to their high cost, complexity, lack of familiarity, and catastrophic rupture due to their brittle nature. For this reason, less than 10% of surgeries in the USA and the UK use ceramic bearing surfaces, although 50% of such surgeries are performed with ceramic bearing surfaces in central Europe [19, 20]. Figure 5.5 shows a ceramic-on-polyethylene (CoP) bearing prosthesis [21].

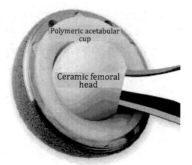

Figure 5.5 Ceramic-on-polymer total hip replacement (femoral head ± zirconia toughened alumina; acetabular cup insert crosslinked polyethylene; metallic acetabular shell with porous coating-cementless fixation). Reprinted from Ref. [25], copyright 2008 Woodhead Publishing Ltd., with permission from Elsevier.

5.4.2 Hard-on-Hard Combinations

In metal-on-metal (MoM) prosthesis, both ball and acetabular liner are made of metal. In case of ceramic-on-ceramic (CoC) prosthesis, the ball and the acetabular liner are made of ceramic material.

So far, MoP is the most commonly used design throughout the world. In India, the emerging trend is MoP for older patients and CoP bearing surfaces for the affordable young patients. In CoP articulation, the ceramic head showed an initial penetration rate (after bedding-in) of 0.5 mm/year. However, after five years, the penetration rate diminishes to 0.1 mm/year, whereas in the case of MoP articulation, it is 0.2 mm/year. MoP also showed more wear and osteolysis compared to that of CoP bearing couple. During the late 1990s, the design of MoM regained popularity among surgeons, especially for younger male patients. However, orthopedic literature published during 2010 cited the problem of early failure in MoM prostheses due to allergy to metallic components. This problem raised concerns among manufacturers, surgeons, and patients worldwide [22]. CoC offered low wear rate without the production of any wear debris, ions, or particles. However, there were a few reports of "squeaking" and catastrophic failure. Biomechanically, CoC design materials were most resistant to wear under *in vivo* condition, which had been documented in 18.5-year-long clinical studies [23]. Hernigou *et al.* [24] found that alumina-on-alumina bearing couple had overall the least osteolysis after long-term clinical study. However, the *survivorship* of alumina–UHMWPE was better than that of alumina-on-alumina for a long term of 15–20 years. Even with cup loosening, osteolysis was found to be lower in the CoC group. At 12-year usage of CoC implant, the average linear wear rate of the alumina-on-alumina group was 0.07 mm/year versus 0.4 mm/year for the Zirconia group. Figure 5.6 shows a comparison of the annual clinical penetration rates (mm/year) and the coefficient of friction values for different bearing combinations [25]. A comparison of merits and limitations of different bearing surfaces is provided in Table 5.1.

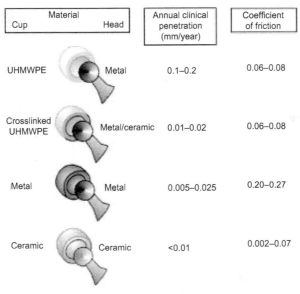

Figure 5.6 Different combinations for bearing surfaces with their annual clinical penetration values and friction coefficients. Reprinted from Ref. [25], copyright 2008 Woodhead Publishing Ltd., with permission from Elsevier.

Table 5.1 Comparative table on merits and limitations of commonly used bearing surfaces in THA worldwide [5, 26, 27]

S. No.	Surfaces	Merits	Limitations
i.	Metal-on-polyethylene (MoP)	• Improved toughness • Long-term clinical survival results over 25 years • Design effects well known	• *In vivo* degradation • Local and systemic toxicity of wear debris • Unsuitable for young patients

(*Continued*)

Table 5.1 (*Continued*)

S. No.	Surfaces	Merits	Limitations
ii.	Metal on highly crosslinked polyethylene (MoXP)	• Increased toughness • Resistance to wear	• Higher cost • Shorter clinical history
iii.	Ceramic on polymer(CoP)	• Abrasive wear resistance • Low friction • Suitable for metal allergic patients • Good long-term results	• Ceramic fracture risk • Limited head sizes • More osteolysis
iv.	Metal-on-metal (MoM) Co–Cr–Mo alloys with finely distributed carbides	• Reduced volumetric wear • History of use over 30 years	• High ion levels • Metal allergy with associated pain for selected patients • Sensitive to abrasion
v.	Ceramic on ceramic (CoC) Fourth-generation (delta) alumina on alumina	• Reduced wear • High abrasion resistance • Highly smooth surface • Low friction • High resistance to third-body wear • Excellent biocompatibility	• Ceramic fracture risk • Depends on the design of the head
vi.	Oxidized zirconium (Oxinium) on HXLPE	• Highly wettable • Good abrasion resistant • Improved toughness • Low friction ceramic surface	• Alloy heads may deform in the case of dislocation

It can be inferred from Table 5.1 that none of the bearing surfaces is an ideal material, each having its own merits and limitations. It was observed that the wear resistance of the implant material played an important role in predicting the risk of revision surgeries. From the clinical point of view, wear has the following major implications:

(1) mechanical failure of the articulation, (2) osteolysis, and (3) loosening [9]. The importance of wear debris dates back to the initial Teflon sockets used by Charnley [13]. As a bearing surface, Teflon has low coefficient of friction, but it has extremely poor wear resistance. The polyethylene wear debris, on the other hand, can stimulate an adverse local host response leading to compromised fixation of the implant, loss of bone stock, and technically difficult bony reconstruction and revision surgery. Submicron polyethylene particles were the dominant type of wear particles present in periprosthetic tissues associated with uncemented hip replacements [23, 28]. Local cascading events led to osteolysis and loosening. Some debris may get transported to the lymphatic system. The relationship between polyethylene wear debris and aseptic loosening has been established beyond doubt in several studies. Sochart [29] examined the above relationship in cemented sockets. At 19.5-year follow-up, the mean total linear wear was observed to be 2.1 mm. Schmalzried *et al.* [30] elucidated the mechanism of biological reaction with polyethylene debris, which eventually led to the loosening of cemented sockets by the examination of autopsy specimens. Dumbleton [31] studied the association between polyethylene wear and osteolysis in THR and suggested that a practical wear-rate threshold of 0.05 mm/year would eliminate osteolysis. Since such wear rates were measured under *in vitro* conditions with highly crosslinked polyethylene, it implied that the said polymer had the potential to eliminate osteolysis and wear-related problems. Continued *in vivo* degradation of prosthetic implant materials was recognized as a major factor diminishing the durability of total joint arthroplasty. Polyethylene wear can be minimized by improving the quality of the material, avoiding the use of large-diameter (greater than 28 mm) femoral heads, improving the design and fabrication of the prosthesis, and reducing the sources of third-body wear [23, 28]. Numerous studies showed a higher rate of wear with the use of metal-backed uncemented cups compared to all-polyethylene cups [32, 33]. The success of a certain bearing couple depends on several other external parameters beyond material-related properties, which decide its effectiveness. The liability of the prosthesis also depends on understanding the biomechanics and the factors that led to its instability, which are discussed in the next section.

5.5 Biomechanics and Implant Stability

The typical load acting on the hip joint and the stability of the implant material influence the longevity of the implant. The hip joint is a typical load-bearing joint subjected to severe impact loading. The biomechanical behavior of an implant decides the influence of its presence on the surrounding tissues and bones [4]. If the modulus of elasticity of the implant is significantly higher than that of the bone, the implant shields the load to be transferred to the bone. It leads to underusage of the implant-loaded region of the bone, consequently leading to gradual weakening of that bone due to the loss of mineral contents. This phenomenon is called *stress shielding*, which might cause or contribute to aseptic loosening of an implant. This happens due to Wolff's law which states that the internal architecture of healthy bone (trabeculae) undergoes adaptive changes depending on the amount of habitual load or stress applied. In other words, remodeling of bone occurs in response to loading via mechano-transduction, a process through which forces or other mechanical signals are converted to biochemical signals in the living cells. The schematic representation of stress shielding is shown in Fig. 5.7 [34].

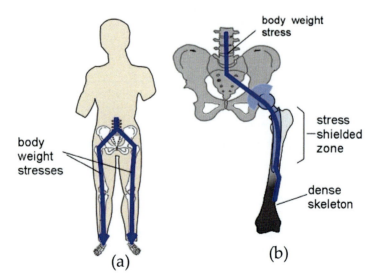

Figure 5.7 (a) Stress flow in a normal healthy human skeleton and (b) changed flow of stress due to stiffer shaft component of total hip prosthesis [34].

Figure 5.7a shows the flow of stress in a healthy human skeleton, which results due to the flow of impulses from the lower back. The stresses flow from the hip to the thigh and finally to the floor through the tibia and ankle. When an artificial hip prosthesis is inserted into the body, the *stiffer* shaft component of the metallic endo-prosthesis takes over the majority of the load stresses leading to the altered distribution of the stresses, as shown in Fig. 5.7b. The altered flow of stress leads to lower load on the upper part of the thigh bone bearing the implant, and thus the bone becomes weaker, which is more susceptible to fracture. Furthermore, the skeleton around the trochanter is overloaded, and it becomes thick and stronger and leads to severe pain in the patient [35].

The stability of the implant is referred to the absence of any visible motion between the loaded bone and the implant [36]. This mechanical quality can be tested surgically, without removing the implant. There are two types of implant stability: (1) primary stability, achieved during surgery, and (2) secondary stability, achieved during postsurgical period. The stability of an implant achieved during the surgical process can be increased during the postoperative period. For example, a press-fit implant ensures primary stability of the cementless components. The primary stability of press-fit acetabular shells can be enhanced by screws or pegs of a combination and by the addition of ribs or flutes, in the femoral stem [4]. The secondary stability of cementless prostheses can be achieved by biological bone ingrowth, and it could be improved by changing the *texture* of the implant surface with porous coating or the calcium phosphate coating. Porous coating can be obtained in different pore sizes in the range of 150 to 1200 µm [4]. A porous coating enlarges the bonding to the surface by three to seven times that of smooth metal, and thus it enhances bone ingrowth. The thickness of the coating seems to be essential, and usually it is in the range of 50 to 110 µm [37]. The adhesive strength of the hydroxyapatite coating on the metal is crucial, and it can potentially induce severe (third-body) wear in the event of early coating degradation and catastrophic failure [38]. It can be understood from the above discussion that the stability of an implant primarily depends on the fixation technique used by the surgeon.

5.6 Fixation Techniques

Factors such as load-bearing time, fixation, cement-initiated wear, and fretting wear typically affect the longevity of the implant. Hence, this section is devoted to the types of fixation techniques and their relative merits and limitations in assisting surgeons and ultimately the prosthesis longevity (survivorship). The fixation techniques that improve the stability of the implant are categorized into three types: (1) cemented fixation, (2) cementless fixation, and (3) hybrid fixation.

5.6.1 Cemented Fixation

Bone cement continues to remain the common form of fixation of THA components due to its less operative and rehabilitation time, comparatively low cost, and usage in patients having irregular medullar cavities as well as in older patients. The cold curing cementing technique is used in cemented THA worldwide. Curing refers to the toughening or hardening of a polymer material by the crosslinking of polymer chains brought about by physical means or chemical additives. However, there are a few grey zones in fully understanding the qualities and properties of cement over time. Postoperative problems such as gradual subsidence of the cemented femoral stem, thermal damage during PMMA polymerization, and the toxic effects of free monomers discourage the use of cement fixation [3–4]. Despite these limitations, the prevailing trend demonstrates that 36% of THRs done in the UK are cemented surgeries, and 16% are of hybrid or reverse hybrid variety [6]. The schematic diagram of cemented fixation is shown in Fig. 5.8 [39].

Nevertheless, inconsistency in the properties of cement continues to challenge surgeons. The sensitivity of cement polymerization against temperature is a well-known confounding variable. Nuno *et al.* [40] measured the transient and residual stresses during the polymerization of PMMA. They used a sub-miniature load cell and a thermocouple setup to measure the radial compressive forces and the temperature, respectively, during the polymerization process and the respective maximum values are found to be 0.6 MPa and 70°C. Fluctuation in the ambient temperature of the operation theatre was implicated as the most important factor. The *dough,*

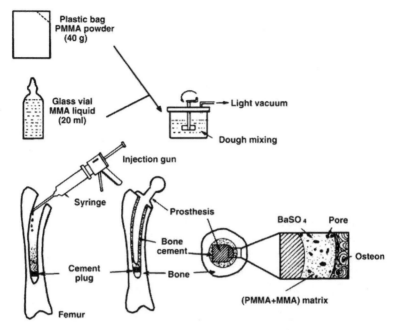

Figure 5.8 Hip joint implantation procedure. Reprinted from Ref. [39], copyright 2007, with permission of Springer.

setting, and *working time* were significantly reduced with an increase of theatre temperature. Farrar and Rose [41] noted that it took about 50% less time for a cement to reach a given viscosity when the temperature changes from 19 to 25°C; thus it is expected to give the surgeon proportionately less time to complete the cementing process. So it was important to maintain the temperature during the cementing process. Antibiotic-eluting cement is a cementing technique for reducing infection-related loosening or revisions, and it can be extrapolated as an advanced use of cement in THA in high-risk groups and revision cases [6]. Heat-stable antibiotics such as Gentamicin, Colistin, Vancomycin, or Cefuroxime may be added. A study identified the preference for the cement in primary hip replacement, considering reports of reduced revision rates with the use of high viscosity cement for aseptic loosening and antibiotic-containing cement in combination with systemic antibiotics for septic loosening [42, 43]. Zamarron *et al.* [44] increased the bioactivity of PMMA bone cement by varying the wollastonite content by soaking the samples in a simulated body fluid (SBF) with an ionic concentration nearly equal to that of human blood plasma. It was reported that the cement with 39 wt.% of wollastonite showed both bioactivity and improved compressive strength. Fukudu *et al.* [45] examined the *in vivo* bone-bonding strength of PMMA bone cement by adding Titania particles with 10, 20, and 30 wt.%. It was reported that the mechanical and thermal properties of Titania–PMMA composite cements were substantially higher than that of commercially available PMMA cement. Pure PMMA is radiolucent and cannot be seen in X-ray films. Makita *et al.* [46] evaluated the changes in mechanical behavior and radiopacity of PMMA by the three-point bending test and computed tomography, respectively, with the addition of various concentrations of $BaSO_4$ such as 10, 20, 30, and 40%. It was found that the radiopacity of PMMA was increased enormously, whereas the mechanical behavior was decreased with an increase of $BaSO_4$. Nien and Huang [47] reinforced PMMA with 0.1, 0.2, 0.27, 0.43, 0.59, and 0.75 wt.% of MWCNT. It was found that the tensile strength and the compressive strength of composites were increased by 18% and 23%, respectively, at the optimum concentration of 0.27 wt.%. It is understood from above discussion that the cemented fixation has some specific advantages such as low cost, less operative time, and less rehabilitation time. Nonetheless, it also has specific disadvantages as discussed. The disadvantages encountered with cemented THA led to an era of cementless THA, which will be elucidated in following subsection.

5.6.2 Cementless Fixation

In uncemented fixation, primary stability is achieved by press-fitting or screwing the components in the bone, whereas secondary stability is achieved by the growth of bone in the implant surfaces. Bone growth is usually achieved by roughening the implant surface through blasting, porous coating, or hydroxyapatite (HA) coating. It helps in bone conservation and a greatly improved long-term *osseointegration* of the implants. Titanium and cobalt–chromium alloys are the two materials primarily used to make cementless total hip prostheses. Titanium is known for excellent strength and resistance to corrosion, while cobalt–chromium alloys exhibit superior corrosion resistance. However, the fatigue resistance of cobalt–chromium alloys is lower than that of titanium. Stainless steel has even lower fatigue strength compared to the other two materials, and hence it is no longer used for cementless hip prostheses [42, 48]. The success of cementless prosthesis is mainly attributed to the correct choice of biomaterials, surface finish and implant geometry, and appropriate surgical techniques. Cementless THR has become very familiar worldwide in the last 20 years. Several long-term studies reported excellent clinical results with a negligible incidence of loosening, thigh pain, osteolysis, or significant stress shielding and, therefore, excellent survivorship approaching 100% after 10 years [3, 4, 42, 48]. The design of the implant and the degree of conformity or match (or mismatch) between the implant and the recipient bone are critical, particularly in cementless fixation. Choosing between cemented and cementless fixation techniques of prosthesis is a continuing debate, each having its own specific advantages and disadvantages. Immediate weight bearing for the function of the operated limb is a problem in cementless designs since it depends on the osseointegration of the implants, which may take several weeks to bond with the recipient bone. This led to the development of a new technique called the *hybrid fixation technique*, which combines the advantages of both the cemented and cementless techniques, as discussed in the following subsection.

5.6.3 Hybrid Fixation

The concept of hybrid fixation partially reduced the problems associated with the use of bone cement, particularly in the younger age group needing THA, where the bone stock is usually high. Hybrid THR involves cemented femoral stem with uncemented acetabular cup (hybrid) or less commonly cemented acetabular implant with uncemented femoral stem (reverse hybrid). Cemented femoral stem is often considered as a gold standard of fixation in THR against which any other designs are compared for clinical trials. Although cementless THR was used for people younger than 65 years, the use of hybrid or reverse hybrid THR has been fairly constant over the years, constituting about 10–17% of all THRs in the UK over the last five years for all age group of male patients [6, 7]. Generally, cementless acetabular cup fixation with cemented femoral fixation is popular for middle-aged people depending on the femoral bone quality. Comparing hybrid versus cementless THR in a long-term study, Kim *et al.* [43, 49] reported the results of 148 hips at a mean of 9.3 years after THA. In their first report (at a mean of 9.3 years), there was no difference in clinical results between the cementless and cemented stems. At the time of follow-up study (at a mean of 17.3 years), there was no difference in clinical results between the hybrid and fully cementless groups. At the time of present review, 40 of 48 acetabular components (83%) in the hybrid group and 80 of 94 acetabular components (85%) in the fully cementless group were intact. Most of the femoral components (98%) in both groups were observed to be intact. A recent report from New Zealand Arthroplasty Registry showed that over a period of 10 years, hybrid THR had the lowest need for revision across all age groups with a 92.94% survivorship [44, 50]. Similarly, the UK Registry also showed a low 3.8% revision rate at 7 years postoperatively.

Figure 5.9 shows the differences among the uncemented THR, cemented THR, and hybrid THR [45, 51]. The cement layer is shown in blue color. Figure 5.9a shows the uncemented technique used for fixing the femoral stem and the acetabular cup. The cemented THR shown in Fig. 5.9b has cemented acetabular cup and femoral stem components. A hybrid THR with cemented femoral component and uncemented acetabular component is shown in Fig. 5.9c.

Although various combinations of materials are used for articulating surfaces using different techniques for both femoral stem and acetabular cup, some critical issues should be addressed for a successful hip replacement surgery in the future. The current limitations encountered in THR are discussed in the next section.

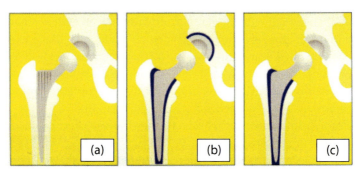

Figure 5.9 (a) Uncemented THR, (b) cemented THR and (c) hybrid THR [45, 51]

5.7 Current Limitations in THR Surgery

The biomechanical failure due to aseptic loosening in hip joint replacement may lead to potential revision of the implant [52]. Failures usually did not occur for 10–15 years in most cases after a well-performed surgery, where the implant selection, implant orientation, and patient selection were done meticulously. Hip prostheses functioned well for up to 20 years in 80% of patients, with a failure rate of nearly 1% per year [52]. Currently, the persistent long-term problems with THRs that remain to be solved include implant wear, osteolysis (periacetabular and/or femoral), and implant loosening, which can be solved by revision surgery [9, 53]. The required surgical skills and cost needed for any kind of revision hip surgery are considerably higher than that of the primary hip replacement surgery due to the loss of bone stock and reconstructive challenges at revision.

5.7.1 Failure Scenario

"Fit and forget" was an essential and most desirable requirement when the implant was in critical applications. Ideally, an implant should serve the lifetime of a patient without any kind of revision. However, the implantation represents a potential assault on the biochemical, physiological, and biomechanical structure of the human body. In general, six types of failure scenarios are identified as potential initiators for revision of hip surgeries, which are briefly discussed below.

5.7.1.1 Accumulated-damage scenario

It was based on the mechanical damage of the material due to repetitive loading, and it depends on the load applied on the material. Although the failure scenario was more relevant to cemented arthroplasty, cementless prostheses can also fail due to mechanical debonding of the implant–bone interface.

5.7.1.2 Destructive-wear scenario

It refers to the mechanical wear of articulating components. The prosthesis is expected to undergo increased wear and then finally lead to failure for younger and highly active patients.

5.7.1.3 Particulate-reaction scenario

It is closely linked with the destructive-wear scenario and depends on the loosening of an implant due to polyethylene wear leading to osteolysis [54, 55].

5.7.1.4 Failed-bonding scenario

It is solely applicable to cementless arthroplasty and is based on the surgical press-fit technique. During rasping, a little gap between the implant and the cavity is always left, which undermines the rigidity of the press-fit. Osteoinductive coating helps to fill the gap and encourages the fixation of an implant.

5.7.1.5 Stress-shielding scenario

It involves only in the stem and the process of load sharing, which depends on the relative stiffness of the stem and the implant material. If the prosthesis is stiffer than the bone, the force will be guided to the prosthesis, with consequential sparing of the bone. The shielding of the bone leads to bone loss. On the other hand, if the bone is stiffer than the implant, proximal osseointegration might fail [56]. Both these phenomena lead to the failure of an implant.

5.7.1.6 Stress-bypass scenario

It is related to the shape of the femoral stem. In the proximal zones, the press-fit is limited due to dissimilar elasticity modulus, which diminishes the proximal stress transfer through the bone leading to bone loss. The prosthesis will subside and find a new position of stability. This enhanced, stable fixation of the implant can be expected only with a tapered femoral stem [57]. A disadvantage of this fixation manifests as thigh pain due to increased hoop stress. To prevent the pain, some authors have advocated a proximal circumferential porous coating over the implant.

From the above-mentioned failure scenarios, it can be concluded that wear particulate generation is one of the prominent material-related problems leading to reduced implant longevity by infection-induced aseptic loosening. Thus, the biocompatibility issues of wear particulates are discussed in the next section.

5.8 Biocompatibility

The success of any material intended to be used for *in vivo* medical applications depends on its biocompatibility. Wear debris has been accepted as the major cause of osteolysis in THA. Concerns have been raised about the biological influence and lifetime effects of wear debris. The local inflammation is reported to be dependent on the particle load (particle size and total volume), aspect ratio, and chemical reactivity. In a normal metal-on-polymer bearing couple, the wear debris is generated from the acetabular cup against femoral head and it can cause serious damage to living tissues. In many patients, the submicron wear particles migrate into the effective joint space and stimulate a foreign-body response resulting in bone loss [58]. It has been noted that the bone around the implant worn away leading to reduced joint function, which is caused by wear-debris-induced osteolysis in the implant [59]. The degradation products of any orthopedic implant include two types of debris: particles and soluble (or ionic) debris. Particulate wear debris includes metals, ceramics, or polymers, which are in the range of nanometer to millimeter. The "metal ions" exist in soluble form in serum protein. The response to implant debris is dominated by local immune activation, e.g., macrophages. However, the effects of systemic increase of metallic and polyethylene wear particles in the liver, spleen, and other tissues of patients have not been associated with remote toxicological or carcinogenic pathology to date [60]. Corrosion of implanted metal by body fluids results the release of unwanted metallic ions with the potential for interference in the process of life. The corrosion resistance is not sufficient enough to suppress the body's reaction to cytotoxic or histotoxic metals or allergic elements such as nickel, leading to rejection of the implant.

Although UHMWPE is a biologically inert material, the presence of a large amount of biologically inactive particulate material can trigger an exceptionally strong macrophage response leading to osteolysis around the implant in general and bone–implant interfaces in particular [61]. Polyethylene particulates induce granulomatous reactions and severe osteolysis when they contact with bone. Therefore, a fully polyethylene cup is not suitable as a cementless prosthesis. Thus, there has been increased focus on the development of alternative techniques/materials for improving the wear resistance of existing materials for total joint replacement to avoid costly revision surgery often necessitated by osteolysis due to wear debris.

5.9 Recent Research and Developments

Currently, UHMWPE is used as a visco-elastic component in acetabular inserts. Due to individual manufacturing processes and subsequent post-manufacturing modifications, the end products have different mechanical properties. Techniques such as reinforcement, irradiation, and hip resurfacing are some of the recent developments in improving the longevity of implants. Reinforcement and irradiation techniques are used in the case of polyethylene, whereas hip resurfacing is a surgical technique used in younger patients using MoM combination. Each technique is briefly explained in the following subsections:

5.9.1 Reinforcement Technique

The selection of reinforcing material used to prepare composites is based on the requirement of the final product in biomedical application. Although a wide variety of reinforcing materials are used to improve the mechanical properties and/or tribological characteristics of UHMWPE, the mechanical properties of a few are compared in Table 5.2.

It can be inferred from Table 5.2 that the presence of reinforcement improved the Young's modulus of UHMWPE. An increase in the toughness of the composites ensures better wear resistance and thus may aid in improving longevity of the implant. It can also be noticed that MWCNTs-reinforced composites showed considerably high enhancement compared to other reinforcing materials.

5.9.2 Highly Crosslinked Polyethylene Acetabular Cup

Highly crosslinked polyethylene (HXLPE) has the potential of considerably reducing wear compared to conventional polyethylene. The reduction in wear debris leads to the reduction in osteolysis and

Table 5.2 Influence of different fillers on mechanical properties of UHMWPE

Filler/blend	Young's modulus	Fracture stress	Yield stress	Percent strain at fracture	Toughness	Researcher
Wollastonite (10%)	Increased from 510 to 630 MPa, i.e., 23.5%	Reduced from 32 to 25 MPa, i.e., 21%	—	—	Reduced from 68 to 40 kJ/m^2, i.e., 41%	Tong et al. [62]
Quartz (20%)	Increased from 1.2 to 1.45 GPa, i.e., 20.8%	Increased from 21 to 24 MPa, i.e., 14.2%	—	—	Increased from 60 to 81 kJ/m^2, i.e., 35%	Xie et al. [63]
HA (sintered and paraffin treated)	Increased from 781 to 1370 MPa, i.e., 75.4%	—	Increased from 6.7 to 11.5 MPa, i.e., 71.6%	—	—	Fang et al. [64]
HA (22.8 wt.%)	Increased from 0.9 to 8 GPa, i.e., 788%	Increased from 27.2 to 28.4 MPa, i.e., 4.4%	—	Decreased from 484 to 358, i.e., 26%	—	Fang et al. [65]
CNF (3 wt.%)	—	Increased from 41 to 60 MPa, i.e., 46.3%	—	—	Increased from 259 to 289 mJ, i.e., 11.5%	Wood et al. [66]
MWCNTs (1 wt.%)	Increased from 0.977 to 1.352 GPa, i.e., 38.3%	—	Increased from 8.27 to 12.38 MPa, i.e., 49.7%	Increased from 4 to 9, i.e., 125%	—	Ruan et al. [67]

(Continued)

Table 5.2 (Continued)

Filler/blend	Young's modulus	Fracture stress	Yield stress	Percent strain at fracture	Toughness	Researcher
MWCNTs (5 wt.%)	Increased from 682 to 1240 MPa, i.e., 81.8%	Reduced from 14.3 to 12.4 MPa, i.e. 9.7%	—	Reduced from 3.9 to 1.4, i.e., 64.1%	—	Bakshi et al. [68]
MWCNTs (3 wt.%)	Increased from 248 to 342 MPa, i.e., 37.9%	Increased from 26.7 to 28.8 MPa, i.e., 7.8%	Increased from 19.9 to 20.7 MPa, i.e., 4.0%	Reduced from 571 to 509, i.e., 10.8%	Reduced from 121 to 115 mJ, i.e., 4.9%	Morlanes et al. [69]
MWCNTs (0.2 to 1 vol.%)	—	—	—	Increased from 275 to 360, i.e, 30.9%	Increased from 50 to 70 J/g, i.e., 40%	Fonseca et al. [70]
MWCNTs (0 to 4 wt.%)	—	Increased from 22 to 27 MPa, i.e., 22.7%	Increased from 20 to 22 MPa, i.e., 10%	—	—	Maksimkin et al. [71]
MWCNTs (2 wt.%)	Increased by 170%	Increased by 93%	Increased by 44%	Increased by 70%	Increased by 176%	Sreekanth and Kanagaraj[72]

aseptic loosening. The clinical usage of HXLPE increases due to an increase of crosslinking by high dosage irradiation and the reduction of oxidative degradation by eliminating free radicals through thermal stabilization [73]. Dorr *et al.* [74] reported the results of retrieval analysis after 5 years of human implantation, where 37 HXLPE and polyethylene liners were compared to each other. It was noticed that the annual linear wear rate of HXLPE was 45% less than that of the conventional polyethylene liner. The prime purpose of irradiation is to enhance the wear resistance of UHMWPE, but it has some deteriorating effects on mechanical properties. The results of a few studies on the wear rate of conventional and crosslinked UHMWPE are given in Table 5.3.

It is observed from Table 5.3 that a higher dose of irradiation led to better wear resistance, leading to increased longevity. However, it is also noticed that the gamma radiation on UHMWPE produced free radicals causing chain scission. When UHMWPE components are irradiated and stored in air, the mechanical properties of components begin to degrade during shelf ageing and the implantation period [86, 87]. UHMWPE implants are also sterilized by ethylene oxide, which eliminates the formation of free radicals. The gamma irradiation of polyethylene results in polymer chains with stable C–C chemical bonds. Coupled with newer postproduction methods, such as various annealing protocols [88, 89], this form of crosslinking greatly improves wear resistance compared to conventional polyethylene, and it helps in reducing the generation of a number of biologically active submicron wear particles. Thus, the clinical results for highly crosslinked UHMWPE have been promising [90]. Another direction of research in countering the oxidative degradation of polyethylene is the development of vitamin E impregnated UHMWPE. Vitamin E (*tocopherol*) is a safe and biocompatible natural antioxidant [91]. The new additive may be effective in decreasing the oxidation and wear rates, but long-term results are awaited as it is an additive with no proven clinical history in joint replacement.

5.9.3 Hip Resurfacing Arthroplasty

Hip resurfacing is yet another arthroplasty technique intended for young patients with an MoM combination [27, 92]. The current

Table 5.3 Influence of irradiation on wear rate/wear factor of UHMWPE.

Irradiation process and dosage	Wear rate/wear factor Virgin	Wear rate/wear factor Irradiated	Type of wear tester	Reference
Lupersol 130	mm^3/year	1.1 mm^3/year	Hip joint simulator	Shen et al. [75]
Gamma irradiation 1000 kGy	17.1 mm^3	3.23 mm^3	Sphere on disc	Oonishi et al. [76]
Gamma irradiation 105 kGy		CoCrMo^{cf} 6.19 ± 4.43 Alumina^{cf} 8.31 ± 3.21 Zirconia^{cf} 0.12 ± 0.14	12-station hip joint simulator	Wang and Schmidig [77]
Gamma irradiation in air 75 kGy	51 mm^3/year	5.62 mm^3/year	Hip joint simulator	Essner [78]
Gamma irradiation 100 kGy	0.649 × 10^{-6} mm^3/N-m	0.449 × 10^{-6} mm^3/N-m	Multidirectional pin-on-disk	Burroughs and Blanchet [79]
50 kGy of electron-beam radiation	0.375 mm^3/250 000 cycles	0.225 mm^3/250 000 cycles	Dual axis simulator	Schwartz et al. [80]
Highly crosslinked (100 kGy)	6.8 × 10^{-7} mm^3/N-m	3.94 × 10^{-7} mm^3/N-m	6-station pin-on-plate	Kang et al. [81]
Highly crosslinked (500 kGy)	15 × 10^{-6} mm^3/N-m	12 × 10^{-6} mm^3/N-m	Ball on disc	Kang and Nho [82]
Gamma irradiation in air 90 kGy	26.8 mm^3/year	17.7 mm^3/year	Knee simulator	Wang et al. [83]
Gamma irradiation 100 kGy	9 g/10^6 cycles	1.8 g/10^6 cycles	Bidirectional pin-on-disc	Oral and Muratoglu [84]
E-beam 72 kGy	14.2 mg/year	4.1 mg/year	Knee simulator	Stoller et al. [85]

cf: counter face

hip resurfacing procedures consist of placing an extremely smooth cobalt–chrome metal cap, which is hollow and shaped like a mushroom over the head of the femur. A highly polished metal cup is placed in the acetabulum (pelvis socket), replacing the articulating surfaces of the patient's hip joint and removing very little proximal femoral bone compared to THR. Hip resurfacing can be considered as a special type of MoM arthroplasty that may allow more range of motion with less risk of dislocation in younger age group. For patients, the benefits of the hip resurfacing technique include large head size, bearing surfaces, and the bone-sparing technique for better kinematics, and thus it becomes a preferred choice for younger-active patients. Birmingham Hip Resurfacing (BHR) is the market leader in resurfacing arthroplasty. It is generally done in moderately advanced symptomatic avascular necrosis of femoral head. Treacy *et al.* [93] reported medium-term results of arthroplasty in 2005 showing 98% overall survival at 5 years in relatively younger, more active adults (average 52.1 year, n = 144). Long-term successful results with BHR have been reported around the globe [94]. In 2009, the Australian Registry reported BHR survivorship at 8 years (95%), which was better than other resurfacing implant survivorship, and also found that resurfacing devices outperformed THR for men under 55 years as well as 55–64 years [95]. Great Britain's Owestry Outcomes Centre's patient registry revealed BHR hip's 10-year survivorship of 95.4% with 98.6% of patients [96]. A study presented at the 2010 annual meeting of the American Academy of Orthopaedic Surgeons discussed several factors important for the successful outcome of resurfacing hip replacements after 8 years of study [97]. But BHR Hip had a revision rate of 11.8% at 7 years, as reported in the UK Joint Registry [6]. The fracture of femur neck has been reported as an important cause of failure due to wrong case selection and compromised surgical techniques, including inadequate cementing or over-pressurization of cement and also osteoporosis [27]. However, it is possible to convert successfully a failed BHR to THR, often leaving the pre-existing acetabular component. Mainly due to its technically demanding nature, limitations, and narrow indications, hip resurfacing has reportedly declined steadily from a peak of 6484 reported procedures in 2006 to 5707 in 2008 and to 2512 in 2010 in some countries [6].

5.10 Conclusion

This chapter reviewed the major progresses made in the field of THR surgeries. The primary aim of THA is to provide a painless, mobile, stable, and durable joint. For addressing this issue, several connected topics have been discussed, beginning with the typical components used in THA and the various biomaterials, articulations, fixation principles, and designs that are used for these components and their evolution over a century, highlighting the major contributions from material researchers and orthopedic surgeons. The currently used biomaterials in THA and their relative merits and demerits have also been discussed. The different fixation techniques, such as cemented, cementless, and hybrid, which help to enhance implant stability, have been discussed along with their merits and limitations. Despite the long-term problems of wear debris and aseptic osteolysis, it has been noted that Charnley's UHMWPE acetabular cup with cemented metallic femoral head is the most preferred articulation among surgeons until recently and that this design is most unlikely to be discarded as an old design. The release of unwanted metallic ions with the potential of interference in the process of life is a cause of concern in MoM designs. It appears that with the known shortcomings of MoP designs and in the backdrop of the rather uncertain future of MoM designs, HXLPE cup and smaller-diameter CoP are slowly emerging as the new *hard-on-soft* articulation of the future, especially for young and affordable patients due to the least risk of wear and osteolysis. The main failure scenarios, the influence of material properties on *in vivo* performances and their selection to avoid undesirable effects such as *stress shielding*, and the need for finding ways and means of achieving optimum implant stability have also been discussed. Recent developments such as reinforced biomaterials, e.g., polyethylene reinforced with MWCNTs, and antioxidants like vitamin E have been found to be encouraging. However, the biocompatibility issues of these new materials are still debatable. Thus, ethical obligations prior to human applications mandate that both high quality *in vitro* and reliable *in vivo* animal studies are to be performed prior to careful clinical case studies followed by stringent human clinical trials. Overall, it can be claimed that the biomaterials used in THA have considerably evolved over the decades from the nascent stages of ivory ball to the present wide variety of advanced composite materials and cross-linked polymers. The success of these biomaterials is likely to significantly improve the longevity of the implants and also the quality of life of the patients. Although certain advanced materials have a lot of promises for the future, nonetheless their successful clinical trials are awaited and such trials must prove their worth and long-term clinical role.

Acknowledgments

Authors duly acknowledge the researchers and working scientists whose contributions are referred to herein.

References

1. Veillette, C. J. H., and Harvey, E. J. (2009) The Information Age: Implications for Orthopaedic Education. *COA Bulletin*. Last modified on 28 August 2011. http://www.coa-aco.org/library/orthopaedic-informatics/the-information-age-implications-for-orthopaedic-education.html.
2. http://evertsmith.com/innovations/ (accessed 08/06/2013).
3. Anish, A. S. C. R., Anoop, G., Joseph, J., and Joseph, J. C. (2010) Analysis of a femur bone implant project report. B.Tech thesis, Department of Mechanical Engineering, College of Engineering, Thiruvanthapuram.
4. Vervest, A. (2005) Zweymüller cementless total hip arthroplasty with two designs of titanium rectangular stem and a titanium threaded cup. A clinical, radiological and DEXA study. PhD thesis, Faculty of Medicine, University of Maastricht. ISBN 90-9018918-1. A.MJ.S. Vervest, Amsterdam, The Netherlands, 2005.
5. Wright, T. M. (2001) Biomaterials and bearing surfaces in total joint arthroplasty: AAOS. *Orthopaedic Knowledge Update,* 50, pp. 57–67.
6. Emsley, D., Newell, C., and Pickford, M. (2009) National Joint Registry for England and Wales, 6th Annual Report. http://www.njr.centre.org.uk (accessed 19/04/2013).
7. Davidson, D., Graves, S., Batten, J., Cumberland, W., Harris, J., and Morgan. (2003) National Joint Replacement Registry, Australian Orthopaedic Association.
8. Brand, R. A., Mont, M. A., and Manring, M. M. (2011) Biographical sketch: Themistocles Gluck (1853–1942) *Clin. Orthop. Relat. Res.* 469(6), pp. 1525–1527.

9. Hallan, G. (2007) Wear, fixation, and revision of total hip prostheses. PhD thesis, Faculty of Medicine, University of Bergen.
10. http://orthopedics.about.com/cs/jointreplacement1/p/austinmoore.htm (accessed 1/7/2014).
11. http://www.maitrise-orthop.com/corpusmaitri/orthopaedic/judet_synth/judet_rencus.shtml (accessed 1/7/2014).
12. Judet, J., and Judet, R. (1950) The use of an artificial femoral head for arthroplasty of the hip joint. *J. Bone Joint Surg.*, 32B, pp. 166–173.
13. Gomez, P. F., and Morcuende, J. A. (2005) Early attempts at hip arthroplasty: 1700s to 1950s. *Iowa Orthop. J.*, 25, pp. 25–29.
14. http://en.wikipedia.org/wiki/John_Charnley (accessed 1/7/2014).
15. Wang, W., Ouyang, Y., and Poh, C. K. (2011) Orthopaedic implant technology: Biomaterials from past to future. *Ann. Acad. Med.*, 40, pp. 237–244.
16. Sandhu, H. S., and Middleton, R. G. (2005) Controversial topics in orthopaedics: Ceramic-on-ceramic. *Ann. R. Coll. Surg. Engl.*, 87, pp. 415–416.
17. Amstutz, H. C., and Grigoris, P. (1996) Metal on metal bearings in hip arthroplasty. *Clin. Orthop. Relat. Res.*, 329, pp. S11–34.
18. Hosseinzadeh, H. R. S., Eajazi, A., and Shahi, A. S. (2012). The bearing surfaces in total hip arthroplasty: Options, material characteristics and selection, In Dr. Samo Fokter (Ed.), *Recent Advances in Arthroplasty*, ISBN 978-953-307-990-5, InTech, pp. 163–210. Available from: http://www.intechopen.com/books/recentadvances-in-arthroplasty/the-bearing-surfaces-in-total-hip-arthroplasty-options-material-characteristics-and-selection.
19. Callaghan, J. J., and Liu, S. S. (2009) Ceramic on crosslinked polyethylene in total hip replacement: Any better than metal on crosslinked polyethylene? *Iowa Orthop. J.*, 29, pp. 1–4.
20. Knight, S. R., Aujla, R., and Biswas, S. P. (2011) Total hip arthroplasty: Over 100 years of operative history. *Orthop. Rev. (Pavia)*, 3(2), pp. e16.
21. http://www.acostawilliams.com/stryker-hip-replacement-recall/ (accessed 10/04/2014)
22. Mckellop, H. A. (2001) Bearing surfaces in total hip replacements: State of the art and future developments. *AAOS Instr. Course Lect.*, 50, pp. 165–179.
23. Hamadouche, M., Boutin, P., and Daussange, J. (2002) Alumina on alumina total hip arthroplasty: A minimum 18.5 year follow-up study. *J. Bone Joint Surg.*, 84, pp. 69–77.

24. Hernigou, P. H., Nogier, A., Poignard, A., and Fillipini, P. (2004) Alumina ceramic against polyethylene: A long term follow up, in *Bioceramics in Joint Arthroplasty, 9th BIOLOX Symposia Proceedings* (Lazennec, J.-Y., and Dietrich, M., eds), pp. 41–42.
25. Revell, P. A. (2008) *Joint Replacement Technology*, Woodhead Publishing, Cambridge, UK.
26. Harris, W. (2001) Wear and periprosthetic osteolysis: The problem. *Clin. Orthop.*, 393, pp. 66–70.
27. Amstutz, H. C., Campbell, P., and Duff, M. J. L. (2007) Metal on metal hip resurfacing: What have we learned? Adult reconstruction: Hip. Section 4. *AAOS Instr. Course Lect.*, 56, pp. 149–169.
28. Jacobs, J. J., Shanbhag, A., Glant, T. T., Blacke, J., and Galante, J. O. (1994) Wear debris in total joint replacements. *J. Am. Acad. Orthop. Surg.*, 2, pp. 212–220.
29. Sochart, D. H. (1999) Relationship of acetabular wear to osteolysis and loosening it total hip arthroplasty. *Clin. Orthop. Rel. Res.*, 363, pp. 135–150.
30. Schmalzried, T. P., Kwong, E. M., Jasty, M., Sedlacek, R. C., Haire, T. C., O'Connor, D. O., Bragdon, C. R., Kabo, J. M., Malcolm, A. J., and Harris, W. H. (1992) The mechanism of loosening of cemented acetabular components in total hip arthroplasty: Analysis of specimens retrieved at autopsy. *Clin. Orthop. Rel. Res.*, 274, pp. 60–78.
31. Dumbleton, A. (2002) Literature review of the association between wear rate and osteolysis in total hip arthroplasty. *J. Arthroplasty*, 17(5), pp. 649–661.
32. Sychterz, C. J., Moon, K. H., Hashimoto, Y., Terefenko, K. M., Engh, C. A., and Bauer, T. W. (1996) Wear of polyethylene cups in total hip arthroplasty. A study of specimens retrieved post mortem. *J. Bone Joint Surg. Am.*, 78, pp. 1193–1200.
33. Nashed, R. S., Becker, D. A., and Gustilo, R. B. (1995) Are cementless acetabular components the cause of excess wear and osteolysis in total hip arthroplasty? *Clin. Orthop. Rel. Res.*, 317, pp. 19–28.
34. http://www.bananarepublican.info/Stress_shielding.htm (accessed 10/04/2014).
35. Surin, H. B. V. (2005) Stress shielding effect of the shaft component. http://www.bananarepublican.info/Stress _shielding.htm (accessed 10/04/2014).
36. Engh, C. A., Sychterz, C., and Engh, C. (1999) Factors affecting femoral bone remodeling after cementless total hip arthroplasty. *J. Arthroplasty*, 14, pp. 637–644.

37. Tsui, Y. C., Doyle, C., and Clyne, T. W. (1998) Plasma sprayed hydroxyapatite coatings on titanium substrates Part 1: Mechanical properties and residual stress levels. *Biomaterials,* 19, pp. 2015–2029.
38. Morscher, E. W., Hefti, A., and Aebi, U. (1998) Severe osteolysis after third-body wear due to hydroxyapatite particles from acetabular cup coating. *J. Bone Joint Surg.,* 80-B, pp. 267–272.
39. Park, J., and Lakes, R. S. (2007) *Biomaterials: An Introduction,* Springer, New York, USA.
40. Nuno, N., Madrala, A., and Plamondon, D. (2008) Measurement of transient and residual stresses during polymerization of bone cement for cemented hip implants. *J. Biomech.,* 41, pp. 2605–2611.
41. Farrar, D. F., and Rose, J. (2001) Rheological properties of PMMA bone cements during curing. *Biomaterials,* 22, pp. 3005–3013.
42. Havelin, L. I. (1995) Hip arthroplasty in Norway 1987–1994. The Norwegian Arthroplasty Register. Bergen, Norway: University of Bergen. http://nrlweb.ihelse.net/Publikasjoner.htm (accessed 20/05/2013).
43. Espehaug, B., Engesaeter, L. B., Vollset, S. E., Havelin, L. I., and Langsland, N. (1997) Antibiotic prophylaxis in total hip arthroplasty. *J. Bone Joint Surg. Br.,* 79, pp. 590–595.
44. Zamarron, D. R., Hernandez, D. A. C., Aragon, L. B., and Lara, W. O. (2009) Mechanical properties and apatite-forming ability of PMMA bone cements. *Mater. Des.,* 30, pp. 3318–3324.
45. Fukuda, C., Goto, K., Imamura, M., Neo, M., and Nakamura, T. (2011) Bone bonding ability and handling properties of a titania–polymethylmethacrylate (PMMA) composite bioactive bone cement modified with a unique PMMA powder. *Acta Biomaterialia,* 7, pp. 3595–3600.
46. Makita, M., Yamakado, K., Nakatsuka, A., Takaki, H., Inaba, T., Oshima, F., and Takeda, H. K. K. (2008) Effects of barium concentration on the radiopacity and biomechanics of bone cement, experimental study. *Radiat. Med.,* 26, pp. 533–538.
47. Nien, Y H., and Huang, C. (2010) The mechanical study of acrylic bone cement reinforced with carbon nanotubes. *Mater. Sci. Eng. B,* 169, pp. 134–137.
48. Swanson, T. V. (2005). The tapered press fit total hip arthroplasty. *J. Arthroplasty,* 20, Supplement 2, pp. 63–67.
49. Kim, Y. K., Kim, J. S., Park, J. W., and Joo, J. H. (2011) Contemporary total hip arthroplasty with and without cement in patients with osteonecrosis of the femoral head. A concise follow-up, at an average

of seventeen years, of a previous report. *J. Bone Joint Surg.*, 93, pp. 1806–1810.

50. Rothwell, A., Taylor, J., and Wright, M. (2009) New Zealand Orthopaedic Association: New Zealand Joint Registry Ten Year Report. http://www.cdhb.govt.nz/njr/reports/A2D65CA3.pdf (accessed 10/05/2013)

51. http://www.orthopaedics.co.uk/boc/patients/total_hip_indications.asp (accessed 10/045/2014)

52. Losina, E., Barrett, J., Mahomed, N. N., Baron, J. A., and Katz, J. N. (2004) Early failures of total hip replacement: Effect of surgeon. *Arthritis Rheum.*, 50, pp. 1338–1343.

53. Yao, J. C. S., Szabo, G., Jacob, J. J., Kuettner, K. E., and Glant, T. T. (1997) Suppression of osteoblast by titanium particle. *J. Bone Joint Surg.*, 79-A, pp. 107–112.

54. Jacobs, J. J., Roebuck, K. A., Archibeck, M., Hallab, N. J., and Giant, T. T. (2001) Osteolysis: Basic science. *Clin. Orthop.*, 393, pp. 71–77.

55. Schmalzried, T. P., and Callaghan, J. J. (1999) Wear in total hip and knee replacements. *J. Bone Joint Surg.*, 81A, pp. 115–136.

56. Engh, C. A., Sychterz, C., and Engh, C. (1999) Factors affecting femoral bone remodeling after cementless total hip arthroplasty. *J. Arthroplasty*, 14, pp. 637–644.

57. Mallory, T. H., Lombardi, A. V., Leith, J. R., Fujita, H., Hartman, J. F., Capps, S. G., Kefauver, C. A., Adams, J. B., and Vorys, G. C. (2002) Why a taper? *J. Bone Joint Surg.*, 84-A (suppl 2), pp. 81–89.

58. Zhu, Y. H., Chiu, K. Y., and Tang, W. M. (2001) Review article: Polyethylene wear and osteolysis in total hip arthroplasty. *J. Orthop. Surg.*, 9, pp. 91–99.

59. Amstutz, H., Campbell, P., Kossovsky, N., and Clarke, I. (1992) Mechanism and clinical significance of wear debris-induced osteolysis. *Clin. Orthop. Rel. Res.*, 276, pp. 7–18.

60. Hallab, N. J., and Jacobs, J. J. (2009) Biologic effects of implant debris. *Bull. NYU Hospital Joint Dis.*, 67(2), pp. 182–188.

61. Peter, D. F., Campbell, P. A., and Amstutz, H. C. (1996) Metal versus polyethylene wear particles in total hip replacements: A review. *Clin. Orthop. Relat. Res.*, 329, pp. S206–S216.

62. Tong, J., Ma, Y., and Jiang, M. (2003) Effects of the wollastonite fiber modification on the sliding wear behavior of the UHMWPE composites. *Wear*, 255, pp. 734–741.

63. Xie, X. L., Tang, C. Y., Chan, K. Y. Y., Wu, X. C., Tsui, C. P., and Cheung, C. Y. (2003) Wear performance of ultrahigh molecular weight polyethylene/quartz composites. *Biomaterials*, 24, pp. 1889–1896.

64. Fang, L., Leng, Y., and Gao, P. (2005) Processing of hydroxyapatite reinforced ultrahigh molecular weight polyethylene for biomedical applications. *Biomaterials*, 26, pp. 3471–3478.

65. Fang, L., Leng, Y., and Gao, P. (2006) Processing and mechanical properties of HA/UHMWPE nanocomposites. *Biomaterials*, 27, pp. 3701–3707.

66. Wood, W. J., Maguire, R. G., and Zhong, W. H. (2011) Improved wear and mechanical properties of UHMWPE–carbon nanofibers composites through an optimized paraffin-assisted melt-mixing process. *Compos.: Part B*, 42, pp. 584–591.

67. Ruan, S. L., Gao, P., Yang, X. G., and Yu, T. X. (2003) Toughening high performance ultrahigh molecular weight polyethylene using multiwalled carbon nanotubes. *Polymer*, 44, pp. 5643–5654.

68. Bakshi, S. R., Tercero, J. E., and Agarwal, A. (2007) Synthesis and characterization of multiwalled carbon nanotube reinforced ultrahigh molecular weight polyethylene composite by electrostatic spraying technique. *Compos.: Part A*, 38, pp. 2493–2499.

69. Morlanes, M. J. M., Castell, P., Nogués, V. M., Martinez, M. T., Alonso, P. J., and Puértolas, J. A. (2011) Effects of gamma-irradiation on UHMWPE/MWNT nanocomposites. *Compos. Sci. Technol.*, 71, pp. 282–288.

70. Fonseca, A., Kanagaraj, S., Oliveira, M. S. A., and Simões, J. A. O. (2011) Enhanced UHMWPE reinforced with MWCNT through mechanical ball-milling. *Defect Diffusion Forum*, 312–315, pp. 1238–1243.

71. Maksimkin, A. V., Kaloshkin, S. D., Kaloshkina, M. S., Gorshenkov, M. V., Tcherdyntsev, V. V., Ergin, K. S., and Shchetinin, I. V. (2012) Ultra-high molecular weight polyethylene reinforced with multi-walled carbon nanotubes: Fabrication method and properties. *J. Alloy. Compd.*, 536, pp. 538–540.

72. Sreekanth, P. S. R., and Kanagaraj, S. (2013) Assessment of surface and bulk properties of UHMWPE/MWCNT nanocomposites using nanoindentation and microtensile testing. *J. Mech. Behavior Biomed. Mat.*, 18, pp. 140–151.

73. Kadar, T., Hallan, G., Aamodt, A., Indrekvam, K. Badawy, M., Skredderstuen, A., Havelin, L. I., Stokke, T., Haugan, K., Espehaug, B., and Furnes, O. (2011) Wear and migration of highly cross-linked and conventional cemented polyethylene cups with cobalt chrome or oxinium femoral heads: A randomized radiostereometric study of 150 patients. *J. Orthop. Res.*, 29(8), pp. 1222–1229.

74. Dorr, L. D., Wan, Z., Shahrdar, C., Sirianni, L., Boutary, M., and Yun, A. (2005) Clinical performance of durasul highly cross-linked

polyethelene acetabular liner for total hip arthroplasty at five years. *J. Bone Joint Surg. Am.*, 87(8), pp. 1816–1821.

75. Shen, F. W., McKellop, H. A., and Salovey, R. (1996) Irradiation of chemically crosslinked ultrahigh molecular weight polyethylene. *J. Polym. Sci. Part B: Polym. Phys.*, 34, pp. 1063–1077.

76. Oonishi, H., Kunot, M., Tsujit, E., and Fujisawa, A. (1997) The optimum dose of gamma radiation–heavy doses to low wear polyethylene in total hip prostheses. *J. Mater. Sci.: Mater. Med.*, 8, pp. 11–18.

77. Wang, A., and Schmidig, G. (2003) Ceramic femoral heads prevent runaway wear for highly crosslinked polyethylene acetabular cups by third-body bone cement particles. *Wear*, 255, pp. 1057–1063.

78. Essner, A. (2005) Hip simulator wear comparison of metal-on-metal, ceramic-on-ceramic and crosslinked UHMWPE bearings. *Wear*, 259, pp. 992–995.

79. Burroughs, B. R., and Blanchet, T. A. (2006) The effect of pre-irradiation vacuum storage on the oxidation and wear of radiation sterilized UHMWPE. *Wear*, 261, pp. 1277–1284.

80. Schwartz, C. J., Bahadur, S., and Mallapragada, S. K. (2007) Effect of crosslinking and Pt-Zr quasicrystal fillers on the mechanical properties and wear resistance of UHMWPE for use in artificial joints. *Wear*, 263, pp. 1072–1080.

81. Kang, L., Galvin, A. l., Thomas, D., Jina, Z., and Fisher, J. (2008) Quantification of the effect of cross-shear on the wear of conventional and highly cross-linked UHMWPE. *J. Biomech.*, 41(2), pp. 340–346.

82. Kang, P. H., and Nho, Y. C. (2001) The effect of γ-irradiation on ultra-high molecular weight polyethylene recrystallized under different cooling conditions. *Radiat. Phys. Chem.*, 60, pp. 79–87.

83. Wang, A., Yau, S. S., Essner, A., Herrera, L., Manley, M., and Dumbleton, J. (2008) A highly crosslinked UHMWPE for CR and PS total knee arthroplasties. *J. Arthroplasty*, 23(4), pp. 559–566.

84. Oral, E., and Muratoglu, O. K. (2007) Radiation cross-linking in ultra-high molecular weight polyethylene for orthopaedic applications. *Nucl. Instr. Meth. Phys. Res. B*, 265, pp. 18–22.

85. Stoller, A. P., Johnson, T. S., Popoola, O. O., Humphrey, M. S., and Blanchard, R. C. (2011) Highly crosslinked polyethylene in posterior-stabilized total knee arthroplasty. *J. Arthroplasty*, 26(3), pp. 483–491.

86. Costa, L., Luda, M. P., Trossarelli, L., Brach, D. P. E. M., Crova, M., and Gallinaro, P. (1998) *In vivo* UHMWPE biodegradation of retrieved prosthesis. *Biomaterials*, 19, pp. 1371–1385.

87. Besong, A. A., Hailey, J. L., Ingham, E., Stone, M., Wroblewski, B. M., and Fisher, J. (1997) A study of the combined effects of shelf ageing following irradiation in air and counterface roughness on the wear of UHMWPE. *Biomed. Mater. Eng.*, 7(1), pp. 59–65.
88. Medel, F. J., Pena, P., Cegonino, J., Barrena, G. E., and Puertolas, J. A. (2007) Comparative fatigue behaviour and toughness of remelted and annealed highly crosslinked polyethylene. *J. Biomed. Mater. Res. B Appl. Biomater.*, 83, pp. 380–390.
89. Kurtz, S. M., Mazzucco, D., Rimnac, C. M., and Schroeder, D. (2006) Anisotropy and oxidative resistance of highly crosslinked UHMWPE after deformation processing by solid-state ram extrusion. *Biomaterials*, 27, pp. 24–34.
90. Rohr, S. M., Li, M. G., Nilsson, K. G., and Nivbrant, B. (2007) Very low wear of non-remelted highly cross-linked polyethylene cups: An RSA study lasting up to 6 years. *Acta Orthop.*, 78, pp. 739–745.
91. Malaika, A. S. (2000) Vitamin E: An effective biological antioxidant for polymer stabilisation. *Polym. Polym. Compos.*, 8, pp. 537–542.
92. McMinn, D. J. (2004) Development of metal/metal hip resurfacing. *Hip Int.*, 13(2), pp. 41–53.
93. Treacy, R. B., McBryde, C. W., and Pynsent, P. B. (2005) Birmingham Hip Resurfacing arthroplasty: A minimum follow up of five years. *J. Bone Joint Surg. Br.*, 87-B, pp. 167–170.
94. Barrack, R. The case for hip resurfacing: A bigger picture. *Current Concepts in Joint Replacement Meeting.* http://www.rediscoveryourgo.com/Product.aspx?Product= BirminghamHip (accessed 16/07/2013).
95. Graves, S. (2009) National Joint Replacement Registry Annual Report. *Australian Orthopaedic Association Adelaide: AOA.* http://www.rediscoveryourgo.com/Product.aspx?Product= BirminghamHip (accessed 16/07/2013).
96. Robinson, E., Richardson, J. B., and Khan, M. Minimum 10 year outcome of Birmingham Hip Resurfacing (BHR), a review of 518 cases from an international register. Oswestry Outcome Centre, Oswestry, UK. http://www.rediscoveryourgo.com/Product.aspx?Product= BirminghamHip (accessed 16/07/2013).
97. Graves, S., Steiger, D. R., Davidson, D., Ryan, P., Miller, L., Stanford, T., and Tomkins, A. (2010) Resurfacing hip replacement: Outcomes at 8 years: An analysis of 12,093 primary procedures. *Proceedings of the American Academy of Orthopaedic Surgeons Annual Meeting, New Orleans LA.* http://www.rediscoveryourgo.com/Product.aspx?Product= BirminghamHip (accessed 16/07/2013).

Index

acetabular cup 177, 192, 195
   cemented 192
   uncemented 191
acetabular liner 180–182
amphiphiles 50, 52, 60, 68, 71, 76
   amide 68
   biocompatible 76
   peptide 73
   prodrug 76
   synthetic 50
amphiphilic molecules 49–51, 68, 76
arthroplasty 201, 204–207, 209
   cemented 193
   low friction torque 179
   total joint 181, 185, 203
artificial dental roots 88
artificial hip prosthesis 177, 187

BGCs *see* bioactive glass ceramics
BHR *see* Birmingham Hip Resurfacing
bioactive behavior 90, 91, 93, 97, 102
bioactive glass ceramics (BGCs) 2, 4, 5, 9
bioactive glasses 1–3, 7, 21, 25, 39, 40, 85–90, 92–102, 104, 106, 108–110, 112, 114
   amorphous 7
   conventional 88, 93
   conventional silicate 100
   conventional sol–gel 99
   ion-substituted 100
   melt-quenched 88
   mesoporous silica 99
   silica-based 7
   silica-based paramagnetic 28
   traditional 96
bioactive materials 3, 4, 19, 20, 25, 39, 72, 91
bioactive molecules 122, 154
biocompatibility 19, 69, 74, 97, 121, 153, 195
biodegradability 40, 121, 123, 126, 153
biodegradable polymers 40, 89, 137
bioglass 1, 3–5, 8, 20, 24, 26, 39, 86, 87, 90, 93, 111–113
   benchmark 3
   conventional 91, 92
   melt-quenched 5
   nanocrystalline 5
bioglass-ceramics 1, 6
biomaterials 1, 2, 19, 39, 49, 121–124, 127, 130, 132, 146–148, 150, 152, 153, 155, 157, 175–177, 202–210
   fiber-based 157
   functional 39
   natural 122, 151
   polymeric 146, 149
   reinforced 202
   synthetic 19, 86
biomedical applications 2, 4, 39, 49, 73, 91, 97, 102, 121, 123, 126, 129, 144, 152, 153, 196
biomineralization 52, 71, 76, 77, 99

Birmingham Hip Resurfacing
 (BHR) 201, 210
block copolymers 50, 65, 67, 73,
 94, 135, 136
bonds 2, 3, 25, 85, 87, 89, 93, 191
 chemical 199
 covalent 70, 122
 hydrogen 50, 127
 polymeric 137
 R–S–S–R 36
bone 4, 19, 25, 52, 71, 72, 74,
 86–89, 93, 131, 145, 146,
 156, 179, 186, 187, 190,
 194, 195
 amputated 74
 autologous 86
 damaged 74
 degenerative 74
 healthy 186
 human 1, 3, 53, 86
 natural 85, 86, 91
 thigh 187
bone cement 176, 188, 191, 206
bone defects 74, 86, 97
bone grafts 74, 85, 101
bone infection 72, 77
bone ingrowth 72, 101, 179, 187
 biological 187
bone regeneration 4, 89, 97
bone replacement 85, 86, 88
bone tissue regeneration 72, 73,
 91, 97
burst tests 146–148

cancer cells 25, 75
cancer chemotherapy 75
cells 85, 89, 90, 98–100, 123–125,
 132, 138, 139, 153, 154
 binding 125
 blood 19
 blood vessel 86
 cancerous 4, 25

 carcinoma MCF-7 66
 fibroblast 156
 living 186
 metastatic MCF-7 75
 osteoblastic 100
 osteosarcoma 75
 stem 39
 sub-miniature load 188
 tumor 128
cementless fixation 179, 188, 190,
 191
cementless prostheses 187, 191,
 193, 195
chitosan 123, 125–129, 156, 157
collagen 52, 75, 121, 125,
 129–131, 156
 bovine 129
 fibrillar 129
 water-insoluble 130
 xenogenic 130
colloidal particles 56, 68
colloidosomes 65, 68
composite materials 96, 97, 203
crystallization 2, 10, 12, 26, 32,
 33, 73, 90, 93, 100, 141,
 142

degradation 38, 39, 88, 101, 109,
 121, 129, 130, 137, 138,
 183, 185
 chemical 52
 controlled congruent 98
 enzymatic 155
 hydrolytic 137
 oxidative 199
 thermal 129, 142
degradation rate 88, 100, 101,
 129, 154
destructive-wear scenario 193,
 194
devices 124, 130, 136, 138, 153
 hemostatic 155

implantable 152
orthopedic fixation 133
prosthetic 153
resorbable fixation 136
resurfacing 201
superconducting quantum interference 15
synthetic biodegradable 133
differential scanning calorimeter (DSC) 13, 14
differential thermal analyzer (DTA) 13, 15
DLS *see* dynamic light scattering
DLS studies 60, 63
drug delivery 1, 4, 39, 49, 65, 66, 73, 123, 126, 131, 138, 153, 154
  controlled 75, 86
  oral 66
  sustained 122
drug delivery systems 52, 75, 91, 122
drug delivery vehicles 67, 153
drugs 66, 67, 70, 72, 76, 142, 154
  anticancer 65, 75
  antimigraine 66, 67
  encapsulated 154
  hydrophobic 67
  lipophilic 69
  liposome-based 67
  low molecular weight 154
  model 74, 75, 153
  non-micellar 66
  non-steroidal anti-inflammatory 68
DSC *see* differential scanning calorimeter
DTA *see* differential thermal analyzer
dynamic light scattering (DLS) 56, 60, 62

EISA process *see* evaporation-induced self-assembly process
electron spin resonance (ESR) 28
electrospinning 124, 125, 131, 138–140, 143, 153, 156, 157
electrospinning process 139, 140, 145
electrospun fibers 141, 145, 153, 154
ESR *see* electron spin resonance
evaporation-induced self-assembly process (EISA process) 94, 95
extracellular matrix 4, 19, 129, 148

fabrication 10, 52, 69, 71, 76, 94, 95, 124, 127, 137–139, 141, 155, 157, 185
failure 146, 149, 176, 193, 194, 201
  biomechanical 193
  catastrophic 182, 187
  mechanical 158, 185
femoral bone 191, 201
femoral head 179, 185, 195, 201, 206
femoral stem 187, 192, 194
  cemented 188, 191
  tapered 194
  uncemented 191
fibers 11, 121, 124, 125, 127, 130, 132, 134, 137–145, 147, 149–151, 154, 157
  collagenous 157
  hydrophobic 126
  natural 121, 122
  rayon 142
  silk 127
  smelted 141

superabsorbent 122
synthetic 122, 132
fibrin 125, 131
fibrinogen 121, 125, 131, 156
fixation 39, 179, 188, 191, 194, 204
    cemented 188, 190
    cemented femoral 191
    cementless acetabular cup 191
    compact 74
    compromised 185
    hybrid 188, 191
    load-bearing orthopedic 134
    stable 194
    uncemented 190
fixation techniques 175, 187–189, 191, 202
fracture 178, 179, 187, 197, 198, 201
free radicals 17, 18, 199
function
    autocorrelation 57
    biological 125
    cellular 90, 123
    joint 195
    strain energy density 146
    tissue 155

gel 7, 68, 70, 94, 98, 131, 138
    bovine serum albumin 69
    carbopol 68
    collagen 75
    dried 94
    hydrated silica 92
    liposome 67
    stimuli-responsive 70
    transparent 69
gelatin 125, 126, 131
glass-ceramics 4, 6, 11–13, 30, 32, 33, 37, 39, 104, 109, 112, 114

ceravital 30
machineable 30
sol–gel-derived 31

HBP *see* human blood plasma
hip 186, 187, 189, 192, 193, 204, 205, 207, 210
    artificial 177
hip arthroplasty 179, 204
hip resurfacing 196, 199, 201, 210
hip surgeries 178, 193
homopolymers 123, 135
human blood plasma (HBP) 19, 20, 190
human body fluids 1, 98
human dermal fibroblast 155
human umbilical cord 157
hyaluronic acid 123, 130, 156
hydrogels 52, 66, 70
    temperature-sensitive 70
hydroxyapatite 3, 11, 19, 30, 31, 53, 72–75, 190, 208
    alendronate-functionalized 74
    nanosized 74
    nanostructured 73
    pure 74, 75
    synthetic 72, 73, 87

implant 25, 86, 87, 123, 133, 158, 176, 177, 179, 180, 185–188, 190, 191, 193–196, 203
    bone 153
    cemented acetabular 191
    load-bearing metallic 77
    metallic 74
    oral 88
    orthopedic 39, 175, 195
    polyethylene-based 181
    press-fit 187
    prosthetic 74

stainless steel 123
tissue-engineered 133
implant materials 4, 5, 32, 131, 133, 157, 176, 184, 194
implant stability 176, 186, 187, 202
infections 128, 155, 178
interactions 1, 3, 15, 16, 19, 28, 32, 33, 55, 58, 60
  atomic-scale 98
  biomolecular 123
  dipole–dipole 36
  electrostatic 50, 62, 129
  fine-scale 98
  hydrogen 97
  intermicellar 62
  intermolecular 76
  interparticle 37, 58
  ionic 74
  non-covalent 49, 50
  super-exchange 15, 27, 32
  super-exchange magnetic 15
  super-exchange type 28
  tail–tail 55
interface 50, 55, 68, 99
  bone–implant 195
  crystal–liquid 5
  implant–bone 193
  stable 25
  water–oil 9
ions 3, 15–17, 20, 26–28, 32, 33, 36, 37, 99, 100, 102, 109, 111, 114, 182
  alkali–alkaline-earth 91
  alkaline earth 33
  antibacterial 36
  charged 3
  iron 15, 16, 26, 28, 29
  paramagnetic 28
  transition metal 17
irradiation 196, 199, 200, 209, 210

light scattering 51, 56, 59
  small-angle 151
  static 56
liposomes 52, 65, 67
loading 148, 149, 180, 186

magnetic resonance imaging (MRI) 4, 39
MBGs *see* mesoporous bioactive glasses
mesoporous bioactive glasses (MBGs) 85, 86, 88, 91–95, 102, 104, 109, 112, 113
micelles 50, 51, 53, 54, 60, 62–66, 71, 94
  anisotropic 64
  block-copolymer-based 66
  cylindrical 55, 60, 65
  ellipsoidal 62
  ionic 71
  linear 64
  nonionic 65
  reverse 55
models 53, 125
  geometric packing 53
  predictive analytical 39
modulus 55, 130, 132, 134, 186, 197, 198
  elasticity 194
molecules 19, 50, 51, 53, 89, 94, 122
  active 72
  biological 50, 69
  bridging 131
  charged 129
  keratin 52
  phospholipid 52
  physisorbed water 113
  protein 130
monomers 54, 122, 132, 134, 136–138
  amphiphilic 51

MRI *see* magnetic resonance imaging
multiwalled carbon nanotubes (MWCNTs) 176, 190, 197, 198, 202, 208
MWCNTs *see* multiwalled carbon nanotubes

nanofibers 10, 73, 123, 140, 144, 145, 154, 156
   polymeric 152, 154, 156
   spun 154
   well-ordered 73
nanoparticles 7, 9, 10, 40, 95
natural materials 89, 121, 125, 132, 150, 151
natural polymers 70, 122, 123, 125, 126, 130, 131, 156
network 2, 3, 102, 107, 124
   branched 64
   covalent 86
   cross-linked 68
   depolymerized 113
   glassy 26, 113
   inorganic 98
   polymerized 94
   pore 101
NMR *see* nuclear magnetic resonance
nuclear magnetic resonance (NMR) 102, 113, 144
nucleation 4, 30, 37, 72, 77

osteolysis 179, 182–185, 191, 193–196, 202, 205–207
oxygen 95
   bridging 22, 23, 103
   non-bridging 22, 103

patients 133, 155, 176–178, 180, 182, 187, 188, 193, 195, 201–203, 206, 208

active 193
male 182, 191
metal allergic 184
young 182, 183, 199
younger-active 201
phases 30, 31, 37, 53, 56, 109, 110, 114
   aqueous 50, 94
   biocompatible inorganic 89
   biomineral 30
   brushite 105, 106
   calcium orthophosphate 105
   continuous hydrocarbon 9
   crystalline 6, 26, 40
   hexagonal 67, 73
   inorganic bioactive 39
   lamellar 64
   magnetite 12
   mineral 6, 72
   nonmagnetite 32
plasma 131, 205
   human blood 19, 190
PLM *see* polarized light microscopy
polarized light microscopy (PLM) 151, 152
polycaprolactone 122, 137, 144
polydioxanone 122, 123, 133, 135, 136, 156
polyethylene 142, 180, 181, 196, 199, 202, 204
   conventional 196, 199
   crosslinked 184, 185, 196, 204, 209, 210
polyglycolide 121, 122, 133, 134, 136, 137
polymer chains 122, 129, 140, 141, 188, 199
polymer fibers 124, 137–139, 141, 143, 157
   natural 121
polymeric fibers 124, 141, 143, 144, 152, 158

synthetic   124, 153
ultrafine   144
polymerization   7, 133, 135, 206
polymers   10, 11, 94, 96, 98, 122,
    124, 125, 132, 135, 137,
    138, 140, 142, 145, 153,
    154, 156, 184, 185
    amorphous   134
    biocompatible   138
    biodegradable   129
    bioresorbable   133
    cross-linked   203
    ideal   124
    polyelectrolyte   140
    protein   125
    sugar   125
    tyrosine-carbonate-based
        experimental   133
polymer solution   138, 139, 141,
    154
polysaccharides   72, 122, 125,
    127, 129
    linear   130
    natural   127
porosity   7, 30, 69, 71, 72, 75, 88,
    90, 94, 99, 113, 144, 156
porous scaffolds   89, 96, 100, 101
prosthesis   179, 180, 182, 185,
    191, 193, 194, 209
    acrylic   179, 180
proteins   19, 33, 35, 67, 72, 99,
    128, 129
    fibrillar   131
    fibrous   129
    sericin   126
    serum   195
    structural   129

rats   19, 66, 101
    calcium-deficient   100
    diabetic   70
    male Sprague–Dawley   70

reactions   3, 4, 9, 98, 99, 148, 151
    biological   185
    chemical   122
    decomposition   13, 14
    granulomatous   195
    heterogeneous phase   11
    hydrothermal   72
    immune system   177
    ion exchange   30
    surface kinetic   33
regeneration   74, 86, 88, 123, 156
    neural stem cell   124
resonance   17, 28
    electron spin   28
    nuclear magnetic   102, 144

SALS *see* small-angle light
    scattering
salts   62
    bile   76
    hydrotropic   62
samples   12–15, 17–19, 22, 24, 35,
    37, 38, 55, 58, 59, 75, 94,
    96, 105, 109, 147, 151
    biological   144
    hydrogenous   58
    plate-shaped   19
SANS *see* small-angle neutron
    scattering
SAXS *see* small-angle X-ray
    scattering
SBF *see* simulated body fluid
SBF solution   24, 33, 109
scaffolds   4, 49, 52, 74, 86, 89–91,
    101, 122, 124, 126, 131,
    138, 155–157
    base   40
    ceramic   90
    macroporous   90
    nanofibrous   126, 156
    protein   52, 71
    three-dimensional   90, 127

tissue-engineered 126
scattering 58
    differential 58
    elastic 58
    interparticle 58
    intraparticle 58
scattering techniques 55, 59
SDA *see* structure-directing agent
shells 52, 68, 76
    atomic 144
    crab 127
    porous 68
    press-fit acetabular 187
    shrimp 127, 128
silica 10, 22, 36, 38, 53, 68, 100, 103, 104
silica-rich layer 92, 93, 113
    hydrated 109
    protonated 92
simulated body fluid (SBF) 19–23, 30, 32–34, 88, 90–92, 99, 100, 109, 112, 113, 149, 190
small-angle light scattering (SALS) 151
small-angle neutron scattering (SANS) 56, 58–60, 62, 63
small-angle X-ray scattering (SAXS) 56, 59
sol–gel glasses 86, 90, 93, 94, 100, 113
sol–gel process 7, 75, 89, 94, 98, 101
sol–gel synthesis 89, 91
species 36, 102, 104, 105, 107, 108, 113, 129
    anionic 112, 113
    chemical 17
specimen 18, 59, 144–146, 149–152, 205
    anisotropic 147
    autopsy 185

dog-bone-shaped 147
dumbbell-shaped 145–147
square 149
square-shaped 145, 149, 150
spinning 124, 125, 131, 138, 141, 153
strain 144–146, 148, 149, 152, 154, 197, 198
shear 150
stress 145, 146, 148, 149, 151, 152, 154, 186, 187
    hoop 194
    linear elastic 146
    mechanical 148
structure-directing agent (SDA) 94
surface 12, 18–20, 33–36, 71, 87, 92, 98, 99, 109, 113, 114, 140, 144, 149, 153, 154, 187, 208
    articulating 176, 180, 181, 192, 201
    biomaterial 18
    ceramic 184
    hybrid 98
    metallic 179
    silica-based pore-wall 105
    wet wound 155
surfactants 9, 50–54, 62, 67, 69, 71, 73, 88, 94
    anionic 60
    biocompatible 65
    nonionic 73, 94
    polymeric 67
surgeries 123, 177, 181, 187
    cemented 188
    hip replacement 192, 193
    revision hip 177, 193
sutures 129, 133, 134, 149, 152
    biodegradable 123
    natural collagen 129
    synthetic absorbable 133

synthesis 4, 7, 10, 11, 49, 53, 73, 85, 93–95, 97, 133, 208
  flame spray 10, 95
  gas phase 10, 95
  hybrid 98
  hydrothermal 10, 11
  sol–gel glass 94
  surfactant-mediated 73
synthetic polymers 40, 68, 72, 122, 124, 125, 131, 132, 152, 156
  biodegradable 132, 133
  resorbable 137

techniques 9–13, 18, 19, 26, 55, 56, 58–60, 65, 86, 89, 95, 138, 142–144, 151, 153, 192, 196
  arthroplasty 199
  bone-sparing 201
  cementless 191
  emulsion 72
  experimental 102
  freeze-drying 96
  surface analytical 33
  uncemented 192
THA *see* total hip arthroplasty
THR *see* total hip replacement
THR
  cemented 192
  hybrid 191, 192
tissue 5, 38, 39, 86, 87, 89, 97, 122–124, 126, 131, 133, 138, 144, 147–149, 151, 153, 155
  animal 130
  bioengineering 121
  biological 156
  damaged 124, 155
  dermal 156
  diseased 25
  elastic 148
  host 3, 25, 89
  hyper-elastic 147
  injured 131
  mineralized 76, 77
  periprosthetic 185
  post-mortem 19
  scar 155
  soft 19, 93
  wound 155
total hip arthroplasty (THA) 176, 179–181, 183, 189, 191, 192, 195, 202–207, 209
total hip replacement (THR) 175–177, 180, 185, 188, 191–193, 201, 204, 207

vesicles 52, 53, 55, 62, 65–67
  bilayer 55
  multilamellar 67
  unilamellar 62

wear 158, 175, 180, 182, 184, 185, 187, 202, 204, 205, 207–210
  cement-initiated 188
  fretting 188
  mean total linear 185
  mechanical 193
  third-body 180, 184, 185, 206

X-ray diffraction (XRD) 11, 99, 102, 142, 151
XRD *see* X-ray diffraction
XRD patterns 11, 20, 30

Young's modulus 144, 145